SpringerBriefs in Applied Sciences and Technology

Computational Intelligence

Series Editor

Janusz Kacprzyk

For further volumes:
http://www.springer.com/series/10618

Ricardo M. F. Martins
Nuno C. C. Lourenço · Nuno C. G. Horta

Generating Analog IC Layouts with LAYGEN II

 Springer

Ricardo M. F. Martins
Instituto de Telecomunicações/Instituto
 Superior Técnico
Lisbon
Portugal

Nuno C. G. Horta
Instituto de Telecomunicações/Instituto
 Superior Técnico
Lisbon
Portugal

Nuno C. C. Lourenço
Instituto de Telecomunicações/Instituto
 Superior Técnico
Lisbon
Portugal

ISSN 2191-530X   ISSN 2191-5318 (electronic)
ISBN 978-3-642-33145-9   ISBN 978-3-642-33146-6 (eBook)
DOI 10.1007/978-3-642-33146-6
Springer Heidelberg New York Dordrecht London

Library of Congress Control Number: 2012946627

© The Author(s) 2013
This work is subject to copyright. All rights are reserved by the Publisher, whether the whole or part of the material is concerned, specifically the rights of translation, reprinting, reuse of illustrations, recitation, broadcasting, reproduction on microfilms or in any other physical way, and transmission or information storage and retrieval, electronic adaptation, computer software, or by similar or dissimilar methodology now known or hereafter developed. Exempted from this legal reservation are brief excerpts in connection with reviews or scholarly analysis or material supplied specifically for the purpose of being entered and executed on a computer system, for exclusive use by the purchaser of the work. Duplication of this publication or parts thereof is permitted only under the provisions of the Copyright Law of the Publisher's location, in its current version, and permission for use must always be obtained from Springer. Permissions for use may be obtained through RightsLink at the Copyright Clearance Center. Violations are liable to prosecution under the respective Copyright Law.
The use of general descriptive names, registered names, trademarks, service marks, etc. in this publication does not imply, even in the absence of a specific statement, that such names are exempt from the relevant protective laws and regulations and therefore free for general use.
While the advice and information in this book are believed to be true and accurate at the date of publication, neither the authors nor the editors nor the publisher can accept any legal responsibility for any errors or omissions that may be made. The publisher makes no warranty, express or implied, with respect to the material contained herein.

Printed on acid-free paper

Springer is part of Springer Science+Business Media (www.springer.com)

To my parents and Nádia
　　　　　　　　　　Ricardo M. F. Martins

To Alina
　　　　　　　　　　Nuno C. C. Lourenço

To Carla, João and Tiago
　　　　　　　　　　Nuno C. G. Horta

Preface

In the last years, the world has observed the increasing complexity of integrated circuits (ICs), strongly triggered by the proliferation of consumer electronic devices. The design of complex system on a chip (SoC) is widespread in multimedia and communication applications, where the analog and mixed-signal (AMS) blocks are integrated together with digital circuitry. However, the analog blocks development cycles are larger when compared to the digital counterpart. The two main reasons identified are the lack of effective computer-aided design (CAD) tools for electronic design automation (EDA), and that analog circuits are being integrated using technologies optimized for digital circuits. Given the economic pressure for high-quality yet cheap electronics and challenging time-to-market constraints, there is an urgent need for CAD tools that increase the analog designers' productivity and improve the quality of resulting ICs.

The work presented in this book belongs to the scientific area of electronic design automation and addresses the automatic generation of analog IC layout. An innovative design automation tool based on template descriptions and on evolutionary computation techniques, LAYGEN II, was developed to validate the proposed approach giving special emphasis to the reusability of expert design knowledge and to the efficiency on retargeting operations. The designer specifies the sized circuit-level structure, the required technology, and, also, provides the technology-independent high-level layout guidelines through an abstract layout description, hence forward called template. The generation proceeds in the traditional way, first placement and then routing. For placement, the topological relations present in the template are mapped to a non-slicing B*-tree layout representation, and the tool automatically merges devices and ensures that the design rules are fulfilled. The router optimization kernel consists of a modified version of the multi-objective evolutionary algorithm (MOEA), NSGA-II, and uses a built-in evaluation engine. The automatic layout generation is here demonstrated using the LAYGEN II tool for two selected typical analog circuit structures, namely, a fully dynamic comparator and a single-ended folded cascode amplifier. The layouts were generated for two design processes, United microelectronics corporation (UMC) 130 nm and Austria microsystems (AMS) 350 nm, and the output provided is a GDSII stream format, a file standard for data exchange of IC layout. Automatic generation processes were performed in less than 5 min, which allow for the designer to quickly obtain a solution. The results were validated using the industrial grade verification tool Calibre® to run design rule check (DRC), layout versus schematic (LVS), and also extraction, in addition post-layout simulations were successfully performed.

This book is organized into seven chapters.

Chapter 1 presents a brief introduction to the area of analog IC design automation, with special emphasis to the automatic layout generation. First, the analog design problem is characterized, then, a well-accepted design flow for analog IC is presented, and finally, LAYGEN II features are outlined.

Chapter 2 starts by addressing the placement problem in EDA, providing a brief overview of the most recent placement tools developed, followed by the presentation of the main references of automatic layout generation tools, and the recent advances in layout-aware analog synthesis approaches. Finally, the available commercial solutions for analog layout automation are outlined.

Chapter 3 gives an overview of the proposed automatic flow for analog IC design, with emphasis on the layout generation task, followed by a general description of the layout generation flow using LAYGEN II. Finally, additional detail about the tool's implementation, inputs, outputs, and interfaces is provided. These interfaces are used by the designer to quickly generate and monitor the automatic generation of the target layout.

Chapter 4 presents the methods used by the Placer to process and place the modules in the floor plan, while following the designer guidelines embedded in the template. First, the general architecture of the Placer is addressed, followed by the description of the high level guidelines present in the template. Finally, the detailed generation procedure for the floor plan, depicting each task implemented in LAYGEN II's template-based Placer is presented.

Chapter 5 covers the general description of the Router architecture, followed by the description of the template information necessary for routing, namely, the connectivity and routing constraints. Then the routing generation procedure is explained, depicting each task implemented in LAYGEN II's optimization-based Router, with emphasis on the evolutionary computational techniques used. Finally, the internal evaluation procedure used to verify if the routing solutions fulfill all the technology design rules and constraints is detailed.

Chapter 6 illustrates the application of the proposed design flow to practical examples. First, a fully dynamic comparator is considered to compare the LAYGEN II results with a hand-made layout using the UMC 130 nm design process. Then, a single-ended folded cascade amplifier is selected to explore the retargetability characteristics of the proposed methodology, using both the UMC 130 nm and the AMS 350 nm design processes.

Chapter 7 summarizes the provided book and supplies the respective conclusion and future work.

Ricardo M. F. Martins
Nuno C. C. Lourenço
Nuno C. G. Horta

Contents

1 Introduction 1
   1.1 Analog IC Design 1
   1.2 The Analog IC Design Automation Flow 3
   1.3 Analog IC Layout Automation 5
   1.4 Conclusions 6
   References 6

2 State of the Art on Analog Layout Automation 9
   2.1 Placement 9
      2.1.1 Layout Constraints 10
      2.1.2 Chip Floorplan Representations 10
      2.1.3 Approaches 15
   2.2 Layout Generation Tools 16
   2.3 Closing the Gap Between Electrical and Physical Design 17
      2.3.1 Layout-Aware Sizing Approaches 19
   2.4 Commercial Tools 21
   2.5 Conclusions 24
   References 25

3 Automatic Layout Generation 29
   3.1 Design Flow Based on Automatic Generation 29
      3.1.1 Sizing Task 31
   3.2 Layout Generation Design Flow 32
   3.3 Tool Architecture 33
      3.3.1 Graphical User Interface 35
      3.3.2 Technology Design Kit 37
      3.3.3 Hierarchical High Level Cell Description 37
   3.4 Conclusions 39
   References 39

4 Placer 41
- 4.1 Placer Architecture 41
- 4.2 Template 42
- 4.3 Template-Based Generation Procedure 43
  - 4.3.1 Instantiation 45
  - 4.3.2 Pre-Processing 45
  - 4.3.3 Post-Processing 50
- 4.4 Conclusions 53
- References 54

5 Router 55
- 5.1 Router Architecture 55
- 5.2 Template 57
- 5.3 Optimization-Based Generation Procedure 60
  - 5.3.1 Multiple Contacts 60
  - 5.3.2 Evolutionary Algorithm 61
  - 5.3.3 Optimization Phases 69
- 5.4 Internal Evaluation Procedure 71
  - 5.4.1 Short Circuit Checker 71
  - 5.4.2 Design Rule Checker 72
  - 5.4.3 Electrical Rule Checker 73
- 5.5 Conclusions 74
- References 75

6 Results 77
- 6.1 Case Study I: Fully-Dynamic Comparator 77
  - 6.1.1 Template 79
  - 6.1.2 Layout Generation 79
  - 6.1.3 Validation 83
- 6.2 Case Study II: Single-Ended Folded Cascode Amplifier 85
  - 6.2.1 Template Hierarchy 85
  - 6.2.2 Layout Generation 86
  - 6.2.3 Retargeting for Different Sizes 87
  - 6.2.4 Retargeting for Different Technology 91
- 6.3 Conclusions 92
- References 92

7 Conclusions and Future Work 95
- 7.1 Conclusions 95
- 7.2 Future Work 96
- References 98

Abbreviations

ASF	Automatically symmetric feasible
AMS	Analog and mixed-signal
BSG	Bounded-sliceline grid
CAD	Computer-aided design
CMOS	Complementary metal oxide semiconductor
DRC	Design rule check
DSP	Digital signal processing
EDA	Electronic design automation
ERC	Electrical-rule check
GA	Genetic algorithm
GUI	Graphical user interface
IC	Integrated circuit
LVS	Layout versus schematic
MOEA	Multi-objective evolutionary algorithm
POF	Pareto optimal front
SA	Simulated annealing
SoC	System-on-a-chip
SP	Sequence pair
TCG	Transitive closure graph
VLSI	Very large scale integration

Chapter 1
Introduction

Abstract This chapter presents a brief introduction to the area of analog integrated circuit (IC) design automation, with special emphasis to the automatic layout generation. First, the analog design problem is characterized, then, a well-accepted design flow for analog IC is presented, and finally, LAYGEN II features are outlined.

Keywords Analog IC design · Automatic layout generation · Computer-aided-design · Electronic design automation

1.1 Analog IC Design

In the last years, the proliferation of consumer electronic devices triggered a huge increase in the microelectronics market. Thanks to the developments made in the last decades in the area of very large scale integration (VLSI) technologies, the designers have the means to build extremely complex multimillion transistor ICs that implement complete systems in a single chip.

In modern telecommunications and multimedia applications the integration of complex systems-on-a-chip (SoC) is a common practice. In these systems, analog or mixed-signal (AMS) blocks are integrated together with digital processors and memory blocks [1, 2]. Even though most functions in today's ICs are implemented using digital or digital signal processing circuitry, analog circuitry are the link between digital circuitry and the continuous-valued external world. The following list enumerates some typical blocks referred as remaining analog forever [3]:

- On the input side of a system, signals from a sensor, microphone or antenna, must be sensed or received, amplified and filtered up to a level that allows digitalization with satisfying signal-to-noise and distortion ratio. Typical

application of these circuits is in sensor interfaces, telecommunication receivers or sound recording;
- On the output side of a system, the signal from digital processing must be reconverted to analog and it has to be strengthened, so that it can drive outside load with low distortion. These circuits are used, e.g., in telecommunication transmitters and loudspeakers;
- Mixed-signal circuits like sample-and-hold, analog-to-digital converters, phase-locked loops and frequency synthesizers. These blocks establish the interface between input/output sides of a system and digital processing parts of a SoC;
- Voltage/current reference circuits and crystal oscillators offer stable and absolute references for the above mentioned circuitry;
- The last block of analog circuits are the extremely high-performance digital circuits. The prime example is state-of-the-art microprocessors that are custom sized like analog circuits, attempting to reach higher speed and lower power consumption.

Despite the fact that analog blocks constitute only a small fraction of the components on mixed-signal IC and SoC designs, the development time of analog blocks is considerably much higher when compared to the development time of the digital ones. The two main reasons identified for the larger development cycle of analog blocks are the lack of effective computer-aided-design (CAD) tools for electronic design automation (EDA), and that analog circuits are being integrated using technologies optimized for digital circuits. For this reason, given the rampant growth of AMS systems, the economic pressure for high-quality yet cheap electronic products and time-to-market constraints, there is an urgent need for CAD tools that increase the analog design productivity and improve the quality of resulting ICs [4].

Today's analog design is supported by circuit simulators, layout editing environments and verification tools, however the design cycle for AMS ICs is still long and error-prone. These circuits suffer from diverse non-idealities and parasitic disturbances that, by not being weighted in the early stages of development, are responsible for design errors and expensive re-design cycles, making them the bottleneck of SoC and mixed-signal ICs design.

In the International Technology Roadmap for Semiconductors [2] is presented the V-Cycle of a design system architecture, which summarizes the differences between analog and digital design automation. The V-Cycle reveals that in the digital domain, EDA is fairly well developed and establish a low-level design process almost fully automated. The main gap in the digital design path of the system design is the tools and methodologies above the behavioral abstraction level. On the other hand, the analog design path reveals that EDA is still in the early stages in almost all stages of the V-Cycle of a system design.

In order to understand the automation of the analog design, the steps in the design flow must be clear. After this brief introduction to the analog IC design, the

1.1 Analog IC Design

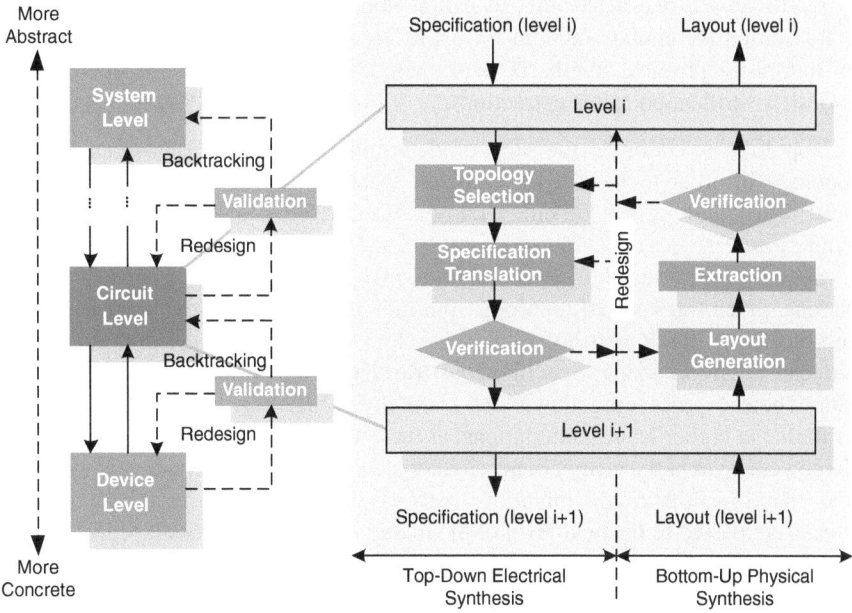

Fig. 1.1 The design flow of analog IC [3]

systematic approach to the analog design automation flow proposed by Gielen and Rutenbar in [3], which intends to ease design automation, is covered in the next section.

1.2 The Analog IC Design Automation Flow

Different analog design automation flows are available in the literature, one that was adopted by many of the works developed in the last decade was the design automation flow introduced by Gielen and Rutenbar in [3]. This design flow for AMS IC circuits is illustrated in Fig. 1.1 and consists of a series of top-down design steps repeated from the system level to the device-level, and bottom-up layout generation and verification. Adopting a hierarchical top-down design methodology is possible to perform system architectural exploration, obtaining a better overall system optimization at a higher abstraction level before starting more detailed implementations at the device level. Thus, problems are found early in the design flow and, as a result, design have a higher chance of first-time success, with fewer or no overall time consuming redesign iterations [1]. The number of hierarchy levels depends on the complexity of the system being handled, and the steps between any two hierarchical levels are:

- The top-down electrical synthesis path includes topology selection, specification translation (or circuit sizing at lowest level) and design verification;
- Bottom-up physical synthesis path includes layout generation and detailed design verification (after extraction).

Topology selection is the step of determining the most appropriate circuit topology in order to meet a set of given specifications at the current hierarchy level. The topology can be either chosen from a set of available topologies, or synthesized.

Specification translation is the step of mapping the high-level block specifications, given the selected topology, into individual specifications for each of the sub-blocks. At the lowest level, the sub blocks are single devices and this task is reduced to circuit sizing. Specifications translation is verified by means of simulation before proceeding down the hierarchy. Since no device-level sizing is available at higher levels, simulations, at these levels, are behavioral. However, at the lowest level in the design hierarchy, the device-level, device sizing is available and, therefore, electrical simulations are used. The specifications for each of the blocks are passed to the next level of the hierarchy and the process is repeated until the top-down flow is completed.

Some recent works based on Pareto optimal fronts (POF) have been very successful exploring the tradeoff during synthesis [5], and already applied at system level sizing. In this approach, a set of non-dominated solutions are generated and the most suitable solution is selected from the POF.

Several CAD tools, settled through the years in the industry, are fundamental to help the designer to successfully complete this task. They are used for IC design editing and evaluation, some of the tools available are: ADiT, Questa, Eldo [6]; HSPICE, nanosim, HSim [7]; Spectre [8]; ngspice [9] and SMASH [10].

Layout generation consists of creating the geometrical layout of the block under design at the lowest level in the design hierarchy, or place and route the layouts of the sub-blocks at higher levels. In the presented design flow, it is important to notice the presence of a detailed verification step over the extraction of the layout. In order to ascend to higher hierarchical levels is necessary that no potential problems are detected at the lowest levels and the layout meet the target requirements. When the topmost level verification is complete, the system is designed and ready for fabrication.

Some CAD tools are available for layout edition, e.g., IC Station Layout [6]; Galaxy Custom Designer LE [7] and Virtuoso Layout Editor [8]. Design rule verification and layout extraction can be performed for example in CALIBRE [6]; Hercules [7] and DIVA, Assura [8].

1.3 Analog IC Layout Automation

The methodology presented in this book focuses on the layout synthesis task of analog integrated circuits, appointed as the critical part of the analog IC design flow [3, 11]. In the last few years a lot of works have emerged from universities, some of them even found their way into commercial EDA tools. Still, the available tools are far from perfect and lots of problems remain unsolved [4]. From an industrial point of view, the application of EDA tools to the analog IC layout synthesis is still far away from being a reality. The onset of more efficient and user-oriented tools is mandatory in order to boost analog designers' productivity and ease this time-consuming task.

Generally the complexity of designing analog circuits' layout is not due to the number of devices, but from the countless interactions between them. Plus, for smaller technology nodes with the increasing complexity of design rules and physical effects these interactions even have a greater impact. Automated analog design tools should concentrate on settling an analog-specific layout synthesis process and take into account all precision matching needs for such designs [2].

LAYGEN [12] was used as starting point for the methodology described in this book. LAYGEN was intended to describe a methodology for automatic analog ICs layout generation, through the introduction of an abstraction level between technological details and the designer guidelines. The abstract layout template captures the designer knowledge independently of technology and specifications. The approach focuses on improving design reusability and retargetability once the template is available, introducing a new level of flexibility by supporting changes on device and modules' specifications.

For the current implementation, design productivity is increased by automatically generating the target layout in a process guided by the designer, and the results validated with a commercial tool, assuring the quality of the solution, which was not done in the previous implementation. Namely, the industrial grade validation of the solutions is performed in Mentor Graphics' Calibre® [6] DRC tool. Technology design rules are strict, which forced reconsidering the whole previous approach. Moreover, the internal module generator generates parametric devices with equivalent quality to the commercial tools' parametric cells.

Fulfilling the restrictive design rules isn't the only objectives of the actual implementation, but also improve the layout quality. The designer provides the high level floorplan and the tool instantiates and places the devices in the layout ensuring that the design rules are strictly respected, automatically abutting devices and including biasing considerations.

A new and more versatile implementation of the router that only requires connectivity to generate DRC clean routing solutions is available, promoting automatic routing generation independently from the floorplan, and consequently allowing topological exploration. Simultaneously, it allows the designer to introduce constraints of symmetry, sensitivity and power nets. In order to evaluate each layout solution without using an external tool, a powerful but lightweight internal validation procedure was developed, to be as reliable as a commercial DRC tool. These are the features of the proposed methodology, and the robustness and retargetability characteristics of LAYGEN II certainly justified its implementation.

LAYGEN II does not intend to replace the designer in the layout generation task, but rather use the designer's knowledge to perform an intelligent pruning of the design space, rapidly providing a solution to be used as a first cut design. Handmade designs are known for their robustness, and the tool provides the means for the designer easily integrate his knowledge and lighten the efforts to accomplish layout design task, abstractly from the intricate technological details.

1.4 Conclusions

Analog IC design automation tools are not keeping up with the challenges created by the technological evolution, while in the digital design several EDA tools and design methodologies are available to help the designers keeping up with the new capabilities offered by the integration technologies. This is one of the reasons why analog design is many technology nodes behind leading-edge digital. Due to the lack of automation, analog designers keep exploring manually the solution space, searching for a solution that fulfills the design specifications. This method causes long design time and, allied to the non-reusable nature of analog IC design, makes it a cumbersome task. After many years of stagnation due to heavy investment in the digital domain, the once-sleepy analog design automation market is now evolving. Simultaneously, LAYGEN II's features were presented, that aim to ease the efforts of analog designers to successfully complete this time-consuming task.

References

1. G.G.E. Gielen, CAD tools for embedded analogue circuits in mixed-signal integrated systems on chip. IEE Proc. Comput. Dig. Tech. **152**(3), 317–332 (2005)
2. International Technology Roadmap for Semiconductors 2009 Edition, http://public.itrs.net/
3. G.G.E. Gielen, R.A. Rutenbar, Computer-aided design of analog and mixed-signal integrated circuits. Proc. IEEE **88**(12), 1825–1852 (2000)
4. R. A. Rutenbar, Analog layout synthesis: what's missing?, in *Proceedings of ACM/SIGDA ISPD*, p. 43 (January 2010)
5. E. Roca, R. Castro-Lopez, F. V. Fernandez, Hierarchical synthesis based on pareto-optimal fronts, in *Proceedings of European Conference on Circuit Theory and Design*, pp. 755–758 (August 2009)
6. Mentor Graphics, http://www.mentor.com
7. Synopsis, http://www.synopsys.com
8. Cadence Design Systems Inc, http://www.cadence.com
9. gEDA Project, http://www.gpleda.org
10. Dolphin Integration, http://www.dolphin.fr

References

11. H. E. Graeb, *Analog Layout Synthesis: A Survey of Topological Approaches,* (Springer, New York, 2010)
12. N. Lourenço, M. Vianello, J. Guilherme, N. Horta, LAYGEN—automatic layout generation of analog ICs from hierarchical template descriptions, in *Proceedings of the Conference on Ph.D. Research in Microelectronics and Electronics*, pp 213–216 (June 2006)

Chapter 2
State of the Art on Analog Layout Automation

Abstract In the past few years, several tools for the automation of the analog integrated circuit (IC) cell and system layout design, with application on both new and reused designs have emerged. Yet, most of the layout design is still handmade, essentially because analog designers want to have total control over the different design options, and also, due to the fact that current fully automated generators of analog IC layouts produce solutions which are not yet competitive with the manually crafted ones. The state-of-the-art on analog layout automation that follows reveals that after many years of stagnation, electronic design automation (EDA) market is evolving, creating more efficient and complementary approaches to the existing tools. The chapter starts by addressing the placement problem in EDA, providing a brief overview of the most recent placement tools developed, followed by the presentation of the main references of automatic layout generation tools, and the recent advances in layout-aware analog synthesis approaches. Finally, the available commercial solutions for analog layout automation are outlined.

Keywords Analog IC design · Automatic layout generation · Chip floorplan representation · Computer-aided-design · Electronic design automation · Layout-aware synthesis

2.1 Placement

Having the devices for the selected topology sized, they must be laid out in the chip, a common analog layout approach is to split the problem into two smaller problems, placement and routing. An automatic placement tool should produce analog device-level layouts similar in density and performance to the high-quality

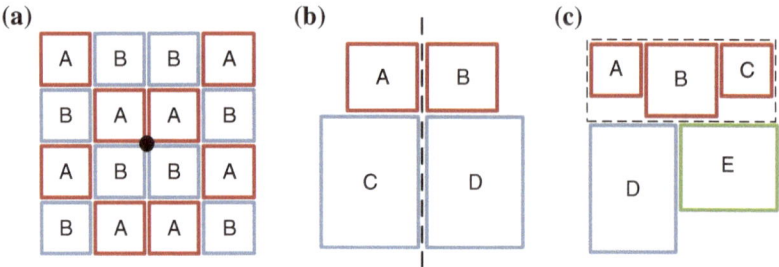

Fig. 2.1 Floorplan constraints **a** Matching (common centroid) **b** Symmetry **c** Proximity (guard ring/well)

manual layouts. In order to achieve this, the capabilities to deal with layout constraints are mandatory.

2.1.1 Layout Constraints

In order to reduce the unwanted impact of parasitic, process variations, different operating conditions and improve the circuit performance, many topological constraints have been introduced into analog placement. The major topological constraints for analog placement are device matching, device symmetry and device proximity [1], as presented in Fig. 2.1. The symmetry constraint restricts devices to a mirrored placement; it is used to offset geometric and electrical issues, and helps reducing the sensitivity to on-die thermal gradients and parasitic mismatches between two identical signal flows.

The matching constraint forces a common gate orientation, common centroid or an interdigitized placement among devices, which improves the beneficial effects of the symmetry constraint by reducing the effect of process-induced mismatches. The proximity constraint limits devices to a specific placement so they can share a common substrate/well region, be surrounded by a common guard ring or be placed close to matched devices. Principally, it decreases the effect of substrate coupling, and also avoids large mismatch and deviations during the fabrication process [1, 2].

2.1.2 Chip Floorplan Representations

Each placement tool has its own strategy of representing the cells, two main different approaches to the chip floorplan representation have been used in the last

2.1 Placement

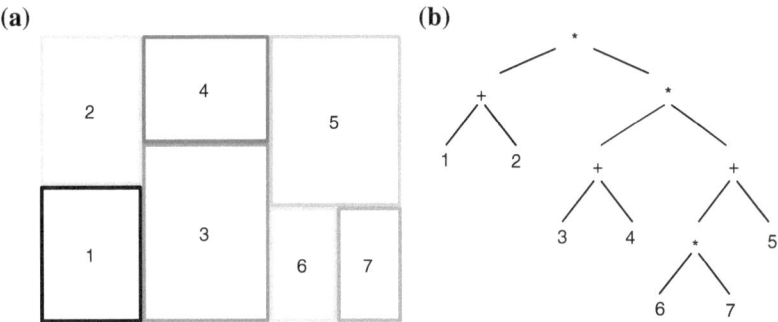

Fig. 2.2 Slicing example **a** Slicing structure, **b** Slicing tree [5]

years [3]: absolute representation (cells are represented by means of absolute coordinates) and topological representation (encoding the positioning relations between any pair of cells), the last one is further classified into slicing or non-slicing representations.

Absolute coordinates is the typical chip floorplan representation used by many computer-aided design (CAD) tools to solve device-level placement problems in analog layout, given the nature of this approach a huge search space is explored. This type of representation allows illegal overlaps during the moves (e.g., translations, changes of orientation), since no restriction is made referring to the relative position of a cell with respect to another cell. To circumvent this situation, a penalty cost term is associated with the total infeasible overlaps, and this penalty must be driven to zero in the, generally, simulated annealing (SA)-based [4] optimization engine [3]. The main disadvantages of using the absolute representation are the high run time, due to the large number of moves necessary to achieve a good layout, once it can generate low-quality and not physically achievable placement solutions, and also, the need of an increased tuning effort due to the difficulty of predicting an appropriate weight for the overlap penalty.

Topological representations trade off a smaller number of moves for more complex-to-build feasible layouts. The first class of topological representation is the slicing model, where cells are organized in sets of slices which recursively bisect the layout horizontally and vertically. The direction and nesting of the slices can be recorded in a slicing tree or, equivalently, in a normalized Polish expression. The typical simulated annealing-based optimizer does not move cells explicitly, as it does when it operates with absolute layout representation. Instead, it alters the relative positions of cells by modifying the slicing tree or normalized Polish expression encoding the layout [1].

Figure 2.2 represents a slicing structure, which is obtained by recursively cut rectangles into smaller ones, the corresponding oriented rooted binary tree (slicing tree) is also presented. The internal nodes marked with "*" represent vertical cuts, and the ones marked with "+" represent horizontal cuts [5]. Since not all the layout topologies have a slicing structure, this representation can degrade the density of the placement solution, which is more noticeable when the cells of a layout are very different in aspect ratio, a common situation in analog circuits. Also, symmetry and matching constraints have to be implemented in the cost function through the use of virtual symmetry axes, which is a less efficient solution.

The weaknesses identified in the slicing model made it a bad choice for placement tools oriented to high-performance analog layout design, culminated in the emergence of several non-slicing topological representations. For these models, the degradation of layout density is no longer a matter of concern [3, 6]. Within the set of non-slicing structures available nowadays one of the most popular is the sequence pair (SP), which encode the "left–right" and "up–down" positioning relations between cells [7], and the solution space can be effectively explored employing SA or genetic algorithms (GA) [8]. Symmetry and device matching constraints can be easily handled, and has a $O(n^2)$ complexity, where n is the number of placeable cells.

The bounded-sliceline grid (BSG) [9] also has a $O(n^2)$ complexity, it uses a meta-grid structure without physical dimensions, but introduces orthogonal relations of "right-of" and "above" unique for each pair of cells. It poses a more intuitive packing than the sequence pair, although for the SP it is proved the existence of a SP that is unconditionally mapped to an optimal packing, while in the BSG this is not always true and its support of symmetry constraints have not been proved yet.

The ordered tree (O-tree) [10] extended the binary tree to the representation of non-slicing structures, presenting a complexity even smaller than in the slicing floorplan. This method was presented to reduce the drawback of redundancies from SP and BSG representations and also, it needs fewer bits to describe the number of blocks than those methods. The run-time for transforming O-tree to its representing placement is linear to the number of blocks ($O(n)$), so one instance of O-tree will map into exactly one placement, no need for extra computation.

An efficient upgraded representation of binary trees, B^*-tree, is also available, offers a $O(n.log(n))$ packing for a binary-tree structure that supports cost evaluation, with no need for additional constraint graphs for cost computation, while the other methods above require them [11]. The correspondence between an admissible placement and its induced B*-tree is one-to-one, so no redundancy, with support for symmetry constraints. In Fig. 2.3 an example of a placement with a symmetry constraint is presented, and their respective representation in O-tree and B*-tree encodings.

More recently, Lin et al. [12] introduced the concept of symmetry island, which keeps modules of the same symmetry group connected to each other. To model a specific placement within a symmetry island a structure based on the B*-tree is used, called Automatically Symmetric Feasible B*-tree (ASF-B*-tree), which also explores symmetry constraints in two dimensions, unlike the other approaches. The principal task of this algorithm is performed on a structure, a Hierarchical B*-tree (HB*-tree), to simultaneously optimize the placement with both symmetry

2.1 Placement

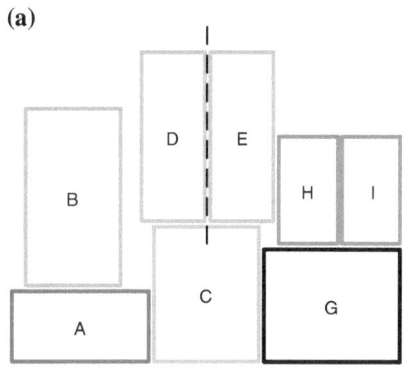
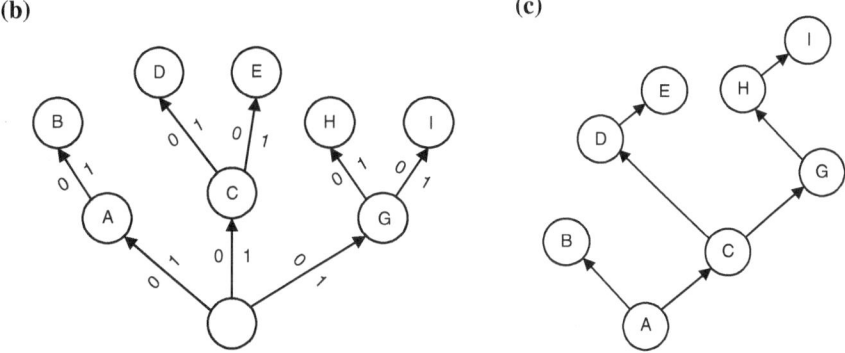

Fig. 2.3 Example of O-tree and B*-tree topological layout representations **a** Placement example **b** O-tree **c** B*-tree [3]

islands and non-symmetry modules, and dynamically update the shape for the devices in a symmetry island. HB*trees are hierarchical oriented, although recent, already proved is high layout quality and runtime efficiency.

The transitive closure graph-based (TCG) [13] representation was proposed to combine the advantages of SP, BSG and B*-tree representations, guaranteeing a unique feasible packing for each representation and that doesn't need to construct additional constraint graphs for the cost evaluation during packing. This approach was quickly replaced for TCG-S [14] derived directly from TCG combined with SP, it presents the fast packing characteristic of SP while maintaining the TCG flexibility to handle placement with special constraints.

In Table 2.1 a summary of the advantages and drawbacks identified for each of the above representations is presented.

Table 2.1 Classification of chip floorplan representations

Representation			Advantages	Drawbacks
Absolute			Easier and quicker-to-build layout configurations, every possible placement can be described; Easiness of modeling positioning constraints.	Slow, solution space infinitely large, requiring a high number of moves; Allow illegal overlaps; Solutions not always physically achievable.
Topological	Slicing	Slicing Tree / Polish Expr. [5]	Smaller solution space; Moves are modifications of the placement code, changing the relative positions of the cells; Cells cannot overlap illegally, which lead to an improved efficiency.	Symmetry and matching constraints are difficult to maintain; Not all layout topologies have a slicing structure, this representation degrade the density of the solutions.
	Non-slicing		All the advantages of the slicing representations; No degradation of layout density.	
		Sequence Pair [7]	A placement configuration can be derived from any encoding; Handles symmetry and device matching constraints.	$O(n^2)$ complexity; $O(n.log(n))$ extra effort for the computation to construct the placement from a SP.
		BSG [9]	More intuitive packing that the sequence pair.	$O(n^2)$ complexity; Cannot always represent the optimal packing for a determined group of cells; Deficient support of constraints.
		O-tree [10]	Smaller complexity and less redundancy than SP and BSG, also fewer bits needed to describe the number of blocks; Transforming O-tree to its representing placement is linear, $O(n)$ effort.	Less flexible than SP and BSG in representation; Tree structure is irregular, and thus some primitive tree operations (e.g., search, insertion) are not efficient.
		B*-tree [11]	Upgrades the O-tree both in processing as in efficiency; Smaller encoding cost; No need for additional constraint graphs;	Less flexible than SP and BSG in representation.
		HB*-tree [12]	Ability to handle symmetry constraints in 2D with Symmetry Islands and ASF-B*-trees; Possibility to combine Symmetry Islands with the rest of the modules.	Less flexible than SP and BSG in representation.
		TCG-S [14]	Combine the advantages of SP, BSG and B*-tree; Best area utilization, faster convergence speed; Flexibility to handle placement with special constraints.	Despite improvements, the evaluation complexity is still quadratic.

2.1.3 Approaches

Over the years different placement tools have explored the advantages of several chip floorplan representations and new ways of treating layout constraints, these tools had been integrated with more or less success in analog synthesis tools. Next are presented some standalone placement tools, developed recently, that somehow present new solutions or significant developments in the classic placement techniques. A more complete background of analog synthesis tools and respective placement approaches, are referred to the Sect. 2.2 of this chapter.

In the area of device-level placement with layout constraints, Krishnamoorthy et al. [15] presented an algorithm based on the exploration of symmetric-feasible SPs, a symmetry group corresponds to a subset of cells which them all share a common symmetry axis. This approach is powered by a $O(G.n.log(n))$ complexity for each code evaluation, where G represents the number of symmetry groups. Koda et al. [16] created an improved method of symmetric placement, obtaining a constraint graph and a set of linear constraint expressions directly from SP while the placement process is accomplished by linear programming.

In the past recent years, developments are being made in hierarchical placement with layout constraints, Lin et al. [17] were the first to present an algorithm for analog placement based on hierarchical module clustering, using HB*-trees. It deals with different constraints simultaneously and hierarchically, this is, if two or more devices are intended to satisfy one or more constraints, they are formed as a cluster, and these clustering constraints can be hierarchically specified to include other clusters. Interesting features for hierarchical symmetry and hierarchical proximity groups that often appear in analog circuits.

In a time dominated by the optimization algorithms, a full deterministic approach arises, Plantage [2], which is based on a hierarchically bounded enumeration of basic building blocks, using B*-trees. This approach is based on the principle that analog circuits show a hierarchical structure, so that hierarchy is used as a bound for the enumeration, aware that a complete enumeration of all possible placements is impracticable. The algorithm begins by generating all placements of the basic modules (leaf nodes of the hierarchy tree), and then the results of the enumerations are combined, guided by the hierarchy tree, until a POF of placements with different aspect ratios for the whole circuit is obtained. Enhanced shape functions are used to store and combine modules efficiently, these functions consist of an ordered set of shapes which are classified through the process by aspect ratio and redundancy, and also modules considered suboptimal are removed for the sake of computational effort.

In a different direction from the other emerging works, Lin et al. [18] proposed a thermal-driven analog placement solution, to simultaneously optimize the placements of "power" and "non-power" (devices which consume much less power than those classified as "power") devices, in an attempt to annihilate thermally-induced mismatches. It is known that the thermal impact from "power" devices can affect the electrical characteristics of the other thermally-sensitive modules, degrading analog and mixed-signal ICs performance. A thermal profile for a given circuit for better thermal matching of the matched devices is established, and the algorithm evolves until the desired thermal profile is achieved. This thermal profile consists essentially in the even distribution of the heat for the whole chip and requires the temperature at each point in the placement area at each iteration.

2.2 Layout Generation Tools

In this section, some of the milestones in the analog layout generation, along with some recent tools, will be reviewed. In the earliest approaches, procedural module generation techniques coded the entire layout of a circuit in a software tool, which would generate the target layout for the parameters attained during sizing. This parametric representation of the layout is fully developed by the designer, either by a procedural language or a graphical user interface (GUI). ALSYN [19] employs fast procedural algorithms that are controlled through a database of structures and attributes. A high-functionality pCell library independent of technologies can be found in [20]. Although fast, these methods lack the flexibility to accommodate wide changes, making the cost of introducing a new design task relatively high and technology migrations may force complete cells redesign.

The use of template approaches, which define the relative position and interconnection of devices, is a common practice. A template-based generation is used by Intellectual Property Reuse-based Analog IC Layout (IPRAIL) [21] to automatically extract the knowledge embedded in an already made layout, and use it for retargeting. Layout retargeting is the process of generating a layout from an existing layout. The main target is to conserve most of the design choices and knowledge of the source design, while migrating it another given technology, update specifications or attempt to optimize the old design [1].

In order to retain the knowledge of the designer but without forcing an implicit definition, LAYGEN [22] uses a template-based approach to guide the layout generation. ALADIN [23] also allow designers to integrate their knowledge into the synthesis process. While ALG [24] uses the same knowledge-based principle, allowing the designer to interact with the tool in different phases, leaving to the discretion of the designer if the final layout is obtained almost full automatically or by designer directives.

Zhang et al. [25] developed a tool that automatically conducts performance-constrained parasitic-aware retargeting and optimization of analog layouts. Performance sensitivities with respect to layout parasitics are first determined, and then the algorithm applies a sensitivity model to control parasitic-related layout geometries, by directly constructing a set of performance constraints subject to maximum performance deviation due to parasitics.

The optimization-based layout generation approaches consist of synthesizing the layout solution using optimization techniques according to some cost

2.2 Layout Generation Tools

Table 2.2 Classification of analog tools based on generation techniques

	Procedural	Template	Optimization
Tools	ALSYN [19], Jingnan [20]	IPRAIL [21], LAYGEN [22], ALADIN [23], ALG [24], Zhang [25]	ILAC [26], KOAN/ANAGRAM II [27], LAYLA [28], Malavasi [29], ALDAC [30]
Advantages	(+) Fast processing; (+) Basic cells	(+) Places modules in a short period of time; (+) Higher abstraction level than procedural; (+) Useful for small adjustments	(+) Higher level of abstraction
Drawbacks	(−) Lack of flexibility, technology migrations force complete cells redesign; high cost of the generation task	(−) Still limits the search space; (−) Designer must add knowledge	(−) Slow; (−) Not always optimal solutions in terms of area and performance

functions, with a higher level of abstraction. The simulated annealing and genetic algorithms are the most common choice for solving analog device-level placement problems, beyond their flexibility in terms of incremental addition of new functionalities; they are relatively easy to implement [6].

In the area of device-level placement with layout constraints there are some main references important to review. ILAC [26] uses simulated annealing operating over a topological slicing tree, used to limit the search space. However, representing the cells by means of absolute coordinates proved to be the most practical solution to implement layout constraints, even though it allows for an infinitely large solution space. This is the approach found in KOAN/ANALGRAM II [27], LAYLA [28], Malavasi et al. [29] and ALDAC [30]. These methods are usually slow and not always produce optimal solutions in terms of area and performance.

In Table 2.2, a classification of the analog tools presented in this section based on generation techniques is presented, and a summary of the advantages of each technique is also highlighted. A summary of the description and functional specifications of the referred tools is presented on Table 2.3, while on Table 2.4 technical specifications and few observations are reported.

2.3 Closing the Gap Between Electrical and Physical Design

In analog IC design, iterations between electrical and physical synthesis to counterbalance layout-induced performance degradations need to be avoided as much as possible [1, 31]. The post-layout performance of a circuit needs to be guaranteed in the presence of layout parasitics, which prevent the circuit from

Table 2.3 Overview of layout generation tools, part I

Layout tool	Years	Specifications			Input/Output
		Description	M[1]	P[2]	
ILAC [26]	1989	Macro-cell Place and Route; placement and routing algorithms inspired by those used for digital design. Not limited to circuits in the input library	✓		In: Netlist, userspecified constraints on cell height; specs. Out: CIF file
KOAN/ ANAGRAM II [27]	1991	Macro-cell Place and Route; uses a pre-defined small module generators database. The fusion of two classical tools, placer KOAN and router ANAGRAM	✓		In: Spice netlist with annotation to control place/route; Out: Magic file
ALSYN [19]	1993	Procedural modules controlled though a user-defined database of rules	✓		In: Circuit netlist; rule sets
LAYLA [28] Malavasi [29]	1995 1996	Similar tools. Macro-cell Place and Route. Constraint-driven layout, the degradation of the performance due to due to interconnect parasitic and device mismatches is weighed, combines this with geometrical optimization	✓	✓	In: Circuit netlist; list of performance specifications
Jingan [20]	2001	Automatic generation and reusability of physical layouts; high-functionality pCells			Procedural layout generation
ALDAC [30]	2002	Generate full-stacked layout modules and performs module placement and local routing. Stacks can be performed either fully-automatically or user controlled	✓		In: Design Rules; ALDAC specific netlist. Out: CIF file
IPRAIL [21]	2003	Automatically creates a template from an existing expertise-embedded layout, and then imposes new device sizes and technology design rules on template	✓		In: CIF file; original and target technology rules. Out: CIF file
LAYGEN [22]	2006	Includes expert knowledge as placement and routing constraints. Designer provides a high level template description and layout is automatically generated	✓		In: Selected template; Technology Design Kit
ALADIN [23]	2006	Designers can develop and maintain technology- and application-independent module generator for relatively complex sub circuits	✓	✓	In: Cells and Netlist; Interative association between them
ALG [24]	2009	User can interact with the tool in each automation step to enhance/polish the layout in order to meet performance specifications. Performance-aware operation provided by a layout adviser tool, YASA	✓	✓	In: Specifications; designer's interaction at different levels
Zhang [25]	2010	Parasitic-aware retargeting; performance sensitivities with respect to parasitics are first determined; automation in a single process without users intervention	✓	✓	In: Existing layout; original and target technology design rules

1—Matching and symmetry constraints; 2—Proximity constraint

reaching the expected performance values. This is usually achieved through time-consuming and unsystematic iterations between the electrical and physical design phases. One possible solution involves the integration of these two different phases, by including layout induced effects into the electrical synthesis phase, or sizing at device-level.

This methodology can be found in recent literature with different designations like parasitic-aware, layout-aware and layout-driven synthesis (or sizing). This is a complex and hard to measure process, since there are geometric requirements whose effects on the resulting parasitics are very specific, so they are never included in the traditional electrical synthesis task. If predicted early in the synthesis process, the overestimation of layout-induced parasitics results in wasted power and area, while underestimation may lead to complete malfunction [6]. Knowing the layout induced effects in the synthesis process ensures that performance of the solution is attained after the layout, and that the area minimization is done more realistically.

2.3.1 Layout-Aware Sizing Approaches

Layout-aware synthesis tools target a design process that avoids time consuming iterations, by bringing layout-related data into the sizing process, even while being aware that layout generation, at each iteration, is an expensive process. In Fig. 2.4, a traditional analog design flow with emphasis on circuit sizing is presented, versus the generalized layout-aware methodology proposed in [35]. Next, some state-of-the-art layout-aware synthesis tools and their different ways of extracting layout-parasitics are presented.

Initially, due to the fast processing of basic cells, procedural-based layout generation techniques were used. Vancorenland et al. [32] used manually derived equations along with a procedural layout generation approach to find a suitable solution. Ranjan et al. [33] generates a parameterized layout using the module specification language system, which consists on a fixed template layout, and when the circuit parameters are provided it produces a physical layout. Then, the extracted parasitic from the layout, along with the passive component values are passed to the precompiled symbolic performance models (symbolic equations in terms of circuit parameters), which predicts the circuit performance at each iteration avoiding numerical simulations.

Without actually generate a layout, Pradhan et al. [34] obtains a Pareto-optimal surface with good spread of points for conflicting performance objectives, and each solution contains the specific layout induced effects. The design space is initially sampled to generate circuit matrix models, which predict circuit performances at each iteration. For a uniform random number of design points, layout samples are generated by a procedural layout generator and device parasitics are modeled by linear regression.

Table 2.4 Overview of layout generation tools, part II

Layout tool	Specifications				Observations
	Placement		Router	Development environment	
	Optimization	Floorplan			
ILAC [26]	S. Annealing	Slicing Tree	Best-first maze search	Pascal	(−) Slicing representation
KOAN/ ANAGRAM II [27]	S. Annealing	Absolute	Re-routing, over-the-device wiring, crosstalk avoidance	C code	(−) High dimensionality of the solution space
ALSYN [19]	Deterministic	Slicing Tree	Maze router with crosstalk avoidance	C code	(+) Combines the concept of easy-to-write rules with fast procedural placement
LAYLA [28] Malavasi [29]	Simulated Annealing	Absolute	Take into account variable wire widths Maze router	C++ code OCTOOLS	(+) Optimize solution quantifying the performance degradation due to impact of parasitic; more optimum solutions found
Jingan [20]	Procedural layout generation			SKILL	Hierarchical parameterized cells
ALDAC [30]	S. Annealing	Absolute	Local routing with two metal layers	C++	Post-layout simulation of multiple layouts
IPRAIL [21]	Linear programming and graph-based methods			–	(−) Undervalues performance
LAYGEN [22]	S. Annealing	B*-tree	Adapts the template routing to the placement	Java	(+) Speeds up retargeting operations
ALADIN [23]	Two-stage: (1) Genetic approach with simulated annealing and half-perimeter routing. (2) Very fast reannealing placement algorithm and global routing			C++, SKILL, Tcl/Tk	(−) Can only handle small or medium size circuits
ALG [24]	Different from custom to automated mode, with global and local routing steps			Java	User may choose the level of automation
Zhang [25]	Mixed-integer nonlinear programming and graph-based methods			C/C++ code	(+) Retargeting with less area and CPU time

2.3 Closing the Gap Between Electrical and Physical Design

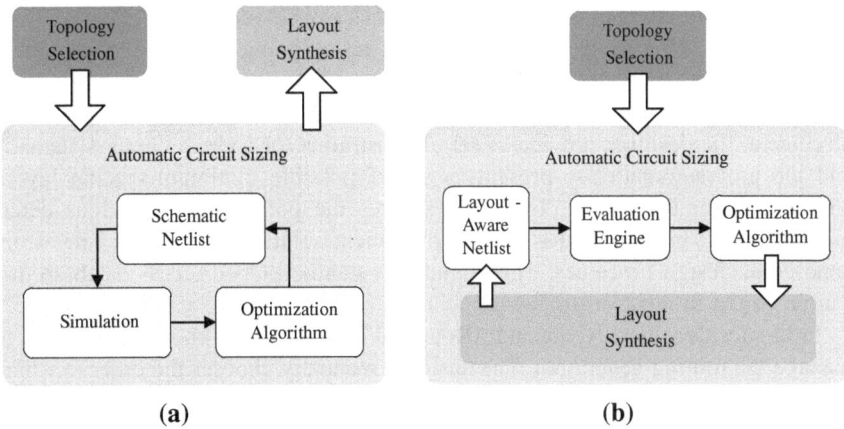

Fig. 2.4 Traditional versus layout-aware circuit sizing flow **a** Traditional analog design flow **b** Layout-aware circuit sizing [35]

Unlike the previous approaches, Castro-Lopez et al. [31] tackles both parasitic-aware and geometrically constrained sizing, which not only includes device parasitic-aware sizing, but also a solution for optimized area and shape. To bypass prohibitively long times for layout generation at each iteration, this approach supports on templates implemented by using the Cadence's pCells technology and SKILL programming. Since the predefined template is supported by a slicing tree and the block placement is fixed, area and shape optimization are obtained by finding the number of fingers of MOS transistors that yield optimal geometric features, or introduced as a constraint to obtain for example, area minimization or a certain aspect ratio. A parasitic estimation without layout generation has been equally implemented, using template sampling techniques and analytical equations.

Recently, Habal et al. [35] ruled out the use of templates given the few degrees of freedom they offer, investigating every possible layout for each device in the circuit using placement algorithm Plantage [2]. The layouts with the best geometric features are kept, and only the final placement selected based on aspect ratio, area and electrical performance is routed. Designer knowledge is supplied by geometric circuit placement and routing constraints, then a deterministic nonlinear optimization algorithm is used for circuit sizing. Table 2.5 shows the summary of the layout aware tools surveyed.

2.4 Commercial Tools

Recently, some commercial solutions have emerged in the analog layout EDA market. Ciranova Helix^{TM} [36] presents as a placement manager supported by a powerful and easy to use graphical user interface (GUI). The designer introduces the system hierarchy and each of the sub-blocks can be added independently from the remainder. This perspective is useful on an on-going system-level specifications translation, since the parasitics from the available blocks and estimated areas can be provided for the designer to optimize the circuit. For the automatic placement, the designer provides a set of constraints for a given circuit schematic and the tool automatically presents a set of possible minimum-spacing layout alternatives for that block. The tool explores the possible combinations deterministically. It produces DRC correct placement solutions for design rules from nano-cmos design processes. The output is a standard OpenAccess database that can be edited in most of the layout editors.

In Mentor Graphics IRoute and ARouter [37] the designer knowledge is used to manage the routing generation. The designer manually chooses the order in which the nets are routed, and the nets with a higher priority are routed more directly. The wires' width and spacing to other nets must be selected, and also, the designer sets the specific conductor to be used in each net and the transition points between layers. The tool provides markers and directions given the set of actual constraints, to interactively help the designer to manually draw the wires.

Calibre® YieldAnalyser [37] integrates process variability analysis using model-based algorithms, that automatically plug layout measurements into yield-related equations to identify the areas of the design that have higher sensitivity to process variations. Critical areas can be mathematically weighted by yield impact information to prioritize and trade-off the issues that have the biggest impact on chip yield. YieldAnalyzer performs critical area analysis on all interconnect layers to identify the areas of a layout with excess vulnerability to random particle defects. The tool runs analysis directly on most of the layout data files, e.g., GDSII, OASIS and OpenAccess design databases, and the information is presented in reports and graphs within the designer's layout environment.

Virtuoso® Layout Suite Family [38] eases the creation and navigation through complex designs, supported by a sturdy multi-window GUI with automatic assistants to aid the designer. These designers' directions guide the physical implementation process while managing multiple levels of design abstractions at device, cell, block, and chip levels, focusing on precision-crafting their designs without sacrificing time to repetitive manual tasks. The suite contains different levels of assistance: basic design-creation and implementation environment; assisted correct-by-construction wire-editing functionalities ensuring real time process-design–rule correctness; captures and drives common hierarchical design intent from schematic editor; and a set of advanced automated finishing tools to optimize the layout and achieve first time successful silicon.

Tanner EDA [39] HiPer DevGen presents a smart generator to accelerate the generation of standard cells. The tool analyzes the netlist and recognizes the current mirrors and differential pairs, and then automatically sends them to the generators. The generated primitives intend to be similar to those handcrafted. Designers have the control over generation options, layout, placement, and routing of these structures. For differential pairs there are multiple options to ensure matching, optimized parasitic, add dummy devices, guard rings, antenna effect

2.4 Commercial Tools

Table 2.5 Overview of layout-aware sizing tools

Tool	Years	Estimation of parasitics	LG[1]	Observations
Vancor [32]	2001	Dimensions of the generated layout are used to calculate analytical models	Yes	(−) Performance and information of parasitics is very limited
Ranjan [33]	2004	Device parasitics extracted from the templates are incorporated into symbolic equation performance models	Yes	(−) Limited to small-signal performances; geometric constraints are not considered
Castro-Lopez [31]	2008	Calculation of the MOS diffusions and areas by analytic equations	No	(−) Storage requirements for the lookup table are exceedingly large
		Geometric methods for transistors, 3-D Analytical and Geometric methods for interconnects and other devices	Yes	(+) Extraction very accurate; performance evaluator is HSPICE (−) Longer simulation times
Pradhan [34]	2009	Sample layouts obtained by procedural generation. Parasitics estimated using polynomial models with the known bias, diffusion areas and perimeters	No	(+) Fast; result is a pareto optimal561 surface inclusive of layout effects; (−) While device parasitics are approximated; geometric constraints and matching are barely considered
Habal [35]	2011	Parasitic coupling capacitances extracted directly by an integral equation field solver without any modeling or approximation	Yes	(+) SPICE simulations for performance evaluation; (−) Slow, complexity of the sizing method increases with the number of devices, routing and parameters

[1]—Layout generation at each iteration

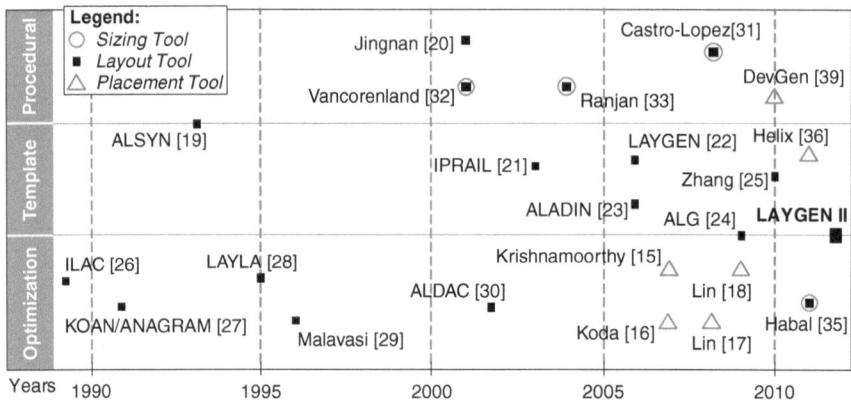

Fig. 2.5 Chronological representation of analog design tools

diodes, etc. For current mirrors there are multiple outputs of different current strengths, options to ensure matching, add dummy devices, share diffusion, multiple finger with options for gate and bulk connections, and adjustments for well proximity effects. To configure a new technology only the design rules are required to have DRC and LVS clean standard cells.

2.5 Conclusions

In this chapter a set of tools applied to analog IC design automation were presented, with emphasis on the layout task. Although much has been accomplished in automatic analog layout generation, the fact is that automatic generators are not yet of standard usage in the industrial design environment. The reviewed approaches are usually limited to some specific circuit or circuit class, and given the large time required to create a new tool or prepare an existing one to support a new circuit, causes analog designers continue to keep on designing the layout manually.

Beyond the efforts made towards the generation of a full automatic layout, capable of competing with expert-made layouts, it is possible to notice that EDA tools are moving in a different direction from two decades ago. There is now a strong attempt to recycle the existing layouts, migrating them to new technologies or optimize the old design. Many of the circuits manufactured today are the same ones developed and implemented years ago, so it is extremely important to take advantage of the knowledge embedded in the layout and follow the advances in the integration technologies, instead of going through all the design process again.

At the same time, layout-aware sizing methodologies are spreading and represent an important part of the future of analog design automation, closing the gap between electrical and physical design for a unified synthesis process. Fast, flexible and as complete as possible layout generation techniques are required to include layout-

2.5 Conclusions

related data into the sizing process, or eventually obtain a final layout simultaneously with the sizing. Most of the layout-aware solutions rely on procedural layout generations, which are known for their difficult reuse and lack of flexibility. The solution of [35] although avoids procedural generations, but at the expense of an increase in the computational time required to complete the automatic flow.

The commercial tools presented, proved that only the approaches that allow for designers to integrate their knowledge into the synthesis process and offer control over the generation, found their way into the EDA market. Most of the available commercial solutions stand out because of the powerful GUIs provided and their characteristics as layout task managers, but lacking on the algorithmic complexity for automatic generation. These tools are used to speed up the manual design task by means of interactive and assisted-edition functionalities.

Figure 2.5 establishes a chronological representation of the tools presented in this chapter, organized by the generation technique used.

From the reviewed approaches, it is possible to notice that floorplan design automation, although far from perfect, is keeping up relatively well with the challenges imposed by new integration technologies. However, the routing task of the proceeding is where the most of the difficulties remain. This is clear when observing the limitations of the current approaches, and the completely lack of routing automation in commercial EDA.

The idea of parameterized model/template is present in the most recent successful approaches. Increasing the designer's active part in generation can't be seen as a drawback, since the inclusion of his knowledge to guide the process increases the layout quality, and consequently the automatic generation tends to present a satisfying solution for the designer. LAYGEN II also allows the designers to integrate their knowledge into the synthesis process, creating an abstraction layer between technological details and the designer guidelines. This design definition is inherently technology independent, allowing changes in circuit's specifications using the same template, which improves the design reusability and focuses on the efficiency of the retargeting operations.

References

1. H. Graeb, *Analog layout synthesis: a survey of topological approaches* (Springer, Berlin, 2010)
2. M. Strasser, M. Eick, H. Gräb, U. Schlichtmann, F.M. Johannes, Deterministic analog circuit placement using hierarchically bounded enumeration and enhanced shape functions in *Proceedings IEEE/ACM International Conference on Computer-Aided Design (ICCAD)*, pp. 306–313, Nov 2008
3. F. Balasa, S.C. Maruvada, K. Krishnamoorthy, On the exploration of the solution space in analog placement with symmetry constraints. IEEE Trans. Comput. Aided Des. Integr. Circuits Syst. **23**(2), 177–191 (2004)
4. B. Suman, P. Kumar, A survey of simulated annealing as a tool for single and multiobjective optimization. J. Oper. Res. Soc. **57**, 1143–1160 (2006)
5. D.F. Wong, C.L. Liu, A new algorithm for floorplan design, in *Proceedings 23th ACM/IEEE Design Automation Conference (DAC)*, pp. 101–107, Jun 1986

6. H. Graeb, F. Balasa, R. Castro-Lopez, Y.-W. Chang, F.V. Fernandez, P.-H. Lin, M. Strasser, Analog layout synthesis—recent advances in topological approaches, in *Proceedings on Design, Automation and Test in Europe (DATE)*, pp. 274–279, 2009
7. F. Balasa, K. Lampaert, Symmetry within the sequence-pair representation in the context of placement for analog design. IEEE Trans. Comput. Aided Des. Integr. Circuits Syst. **19**(7), 721–731(HJul. 2000)
8. A.E. Eiben, J.E. Smith, *Introduction to Evolutionary Computing* (Springer, Berlin, 2003)
9. S. Nakatake, K. Fujiyoshi, H. Murata, Y. Kajitani, Module packing based on the BSG-structure and IC layout applications. IEEE Trans. Comput. Aided Des. Integr. Circuits Syst. **17**, 519–530 (1998)
10. P.-N. Guo, C.-K. Cheng, T. Yoshimura, An O-tree representation of nonslicing floorplan and its applications, in *Proceedings 36th ACM/IEEE Design Automation Conference (DAC)*, pp. 268–273, 1999
11. Y.-C. Chang, Y.-W. Chang, G.-M. Wu, S.-W. Wu, B*-trees: A new representation for nonslicing floorplans, in *Proceedings 37th ACM/IEEE Design Automation Conference (DAC)*, pp. 458–463, 2000
12. P.-H. Lin, S.-C. Lin "Analog placement based on novel symmetry-island formulation, in *Proceedings 44th Design Automation Conference (DAC)*, pp. 465–470, 2007
13. L. Jai-Ming, C. Yao-Wen, TCG: a transitive closure graph-based representation for non-slicing floorplans, in *Proceedings 38th Design Automation Conference (DAC)*, pp. 764–769, 2001
14. L. Lin, Y.-W. Chang, TCG-S orthogonal coupling of P-admissible representations for general floorplans. IEEE Trans. Comput. Aided Des. Integr. Circuits Syst. **23**(5), 968–980 (2004)
15. K. Krishnamoorthy, S. C. Maruvada, F. Balasa, Topological placement with multiple symmetry groups of devices for analog layout design, in *Proceedings IEEE International Symposium on Circuits and Systems (ISCAS)*, pp. 2032–2035, May 2007
16. S. Koda, C. Kodama, K. Fujiyoshi, Linear programming-based cell placement with symmetry constraints for analog IC layout. IEEE Trans. Comput. Aided Des. **26**(4), 659–668 (2007)
17. P.-H. Lin, S.-C. Lin, Analog placement based on hierarchical module clustering, in *Proceedings 45th ACM/IEEE Design Automation Conference (DAC)*, pp. 50–55, June 2008
18. P.-H. Lin, H. Zhang, M. Wong, Y.-W. Chang, Thermal-driven analog placement considering device matching, in *Proceedings 46th ACM/IEEE Design Automation Conference (DAC)*, pp. 593–598, Jul 2009
19. V. Meyer, ALSYN: Flexible rule-based layout synthesis for analog ICs. IEEE J. Solid State Circuits **28**(3), 261–268 (1993)
20. X. Jingnan, J. Vital, N. Horta, A SKILLTM—based library for retargetable embedded analog cores, in *Procedings on Design, Automation and Test in Europe (DATE)*, pp. 768–769, Mar 2001
21. N. Jangkrajarng, S. Bhattacharya, R. Hartono, C. Shi, IPRAIL—Intellectual property reuse-based analog IC layout automation. Integr. VLSI J. **36**(4), 237–262 (2003)
22. N. Lourenço, M. Vianello, J. Guilherme, N. Horta, LAYGEN—Automatic layout generation of analog ics from hierarchical template descriptions, in *Proceedings Conference on Ph.D. Research in Microelectronics and Electronics (PRIME)*, pp. 213–216, Jun 2006
23. L. Zhang, U. Kleine, Y. Jiang, An automated design tool for analog layouts, IEEE Trans. Very Large Scale Integr. (VLSI) Syst. **14**(8), 881–894 (Aug 2006)
24. Y. Yilmaz, G. Dundar, Analog layout generator for CMOS circuits. IEEE Trans. Comput. Aided Des. Integr. Circuits Syst. **28**(1), 32–45 (2009)
25. L. Zhang, Z. Liu, "A performance-constrained template-based layout retargeting algorithm for analog integrated circuits, in *Proceedings 47th ACM/IEEE Design Automation Conference (DAC)*, pp. 293–298, Jan 2010
26. J. Rijmenants, J. Litsios, T. Schwarz, M. Degrauwe, ILAC: An automated layout tool for analog CMOS circuits. IEEE J. Solid State Circuits **24**(2), 417–425 (1989)
27. J.M. Cohn, D.J. Garrod, R.A. Rutenbar, L.R. Carley, KOAN/ANAGRAM II: New tools for device-level analog placement and routing, IEEE J. Solid State Circuits **26**(3) 330–342, Mar 1991
28. K. Lampaert, G. Gielen, W. Sansen, A performance-driven placement tool for analog integrated circuits. IEEE J. Solid State Circuits **30**(7), 773–780 (1995)

References

29. E. Malavasi, E. Charbon, E. Felt, A. Sangiovanni-Vincentelli, Automation of IC layout with analog constraints. IEEE Trans. Comput. Aided Des. Integr. Circuits Syst. **15**(8), 923–942 (1996)
30. P. Khademsameni, M. Syrzycki, A tool for automated analog CMOS layout module generation and placement, in *Proceedings IEEE Canadian Conference on Electrical and Computer Engineering*, vol. 1, pp. 416–421, May 2002
31. R. Castro-Lopez, O. Guerra, E. Roca, F. Fernandez, An integrated layout-synthesis approach for analog ICs. IEEE Trans. Comput. Aided Des. Integr. Circuits Syst. **27**(7), 1179–1189 (2008)
32. P. Vancorenland, G. V. der Plas, M. Steyaert, G. Gielen, W. Sansen, A layout-aware synthesis methodology for RF circuits, in *Proceedings IEEE/ACM International Conference on Computer-Aided Design (ICCAD)*, pp. 358–362, Nov 2001
33. M. Ranjan, W. Verhaegen, A. Agarwal, H. Sampath, R. Vemuri, G. Gielen, Fast, layout inclusive analog circuit synthesis using pre-compiled parasitic-aware symbolic performance models, in *Proceedings Design Automation Conference and Test in Europe Conference (DATE)*, vol. 1, pp. 604–609, Feb 2004
34. Pradhan, R. Vemuri, Efficient synthesis of a uniformly spread layout aware Pareto surface for analog circuits, in *Proceedings 22nd International Conference on VLSI Design*, pp. 131–136, Jan 2009
35. H. Habal, H. Graeb, Constraint-based layout-driven sizing of analog circuits. IEEE Trans. Comput. Aided Des. Integr. Circuits Syst. **30**(8), 1089–1102 (2011)
36. "Ciranova," http://www.ciranova.com/
37. Mentor Graphics, http://www.mentor.com
38. Cadence Design Systems Inc, http://www.cadence.com
39. Tanner EDA, http://www.tannereda.com/

Chapter 3
Automatic Layout Generation

Abstract This chapter gives an overview of the proposed automatic flow for analog integrated circuit (IC) design, with emphasis on the layout generation task, followed by a general description of the layout generation flow using LAYGEN II. Finally, additional detail about the tool's implementation, inputs, outputs and interfaces is provided. These interfaces are used by the designer to quickly generate and monitor the automatic generation of the target layout.

Keywords Analog IC design · Automatic layout generation · Electronic design automation · Evolutionary computation · Layout retargeting · Technology migration

3.1 Design Flow Based on Automatic Generation

It is acknowledged that each designer/company has its own layout style but often this style is very regular for a large number of applications, even with some specifications or technological changes. The design guidelines for most common cells are kept the same. For simple cells, parametric generators are a valid solution to implement these guidelines, however, parametric generators are specific to a technology making them difficult to reuse. In addition, for cells that are more complex, the development of effective parametric generators has proven ineffective either on design-time or on design-reusability.

In order to cope with these limitations and increase design efficiency, LAYGEN II stores these design regularities in a layout meta-description that is independent of technology and sizing (obtained during the specification translation task). The template, together with LAYGEN II and a set of parametric module generators at

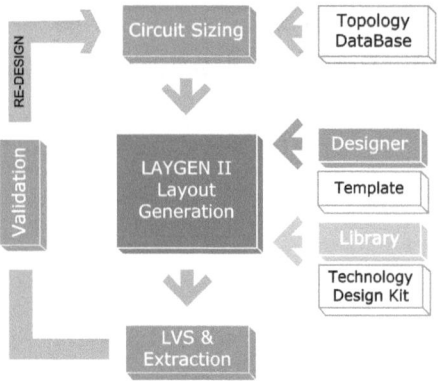

Fig. 3.1 Analog IC design flow: Closed loop

device-level provide the designer with a technological and specification independent way of defining some of the most commonly used cells.

The generation starts from the coarse layout definition described in the template, and finishes at the optimized target layout. In addition, since technological details are treated automatically by LAYGEN II, the designer's efforts are focused on difficult layout issues and not on the technological dimensions details. This way, designer's efforts are better-used, increasing design productivity, efficiency and reusability.

As stated, this methodology considers an abstraction level between the designer expertise and the technology details, which introduces some changes in the typical IC design flow. The proposed automatic design flow for analog ICs is shown in Fig. 3.1. This work addresses how the guidelines are provided to LAYGEN II and the methods used to automatically generate the target layout. Although the tool is being developed to be integrated in the bottom-up physical synthesis path of an analog design automation process, it can also be used as a standalone tool by designers in the layout generation task.

The flow is triggered by a specification translation at system level or a sizing task at circuit level. When a set of device sizes is obtained for the desired specifications this information is included in the selected template, which encompasses the guidelines of the designer, then, layout is automatically generated by LAYGEN II. The additional layout versus schematic (LVS) and extraction steps are necessary to run post-layout simulation and validate the generated design. If the specifications are not met re-design is necessary. The LVS validation verifies if a particular IC layout corresponds to the electrical schematic, this step is only done if a successful design rule check (DRC), for the automatically generated layout, was previously obtained. DRC validation is embedded in the layout generation task, as it will be shown later in LAYGEN II's architecture. Although outside the scope of this work, the sizing task is also briefly covered in the following subsection.

3.1 Design Flow Based on Automatic Generation

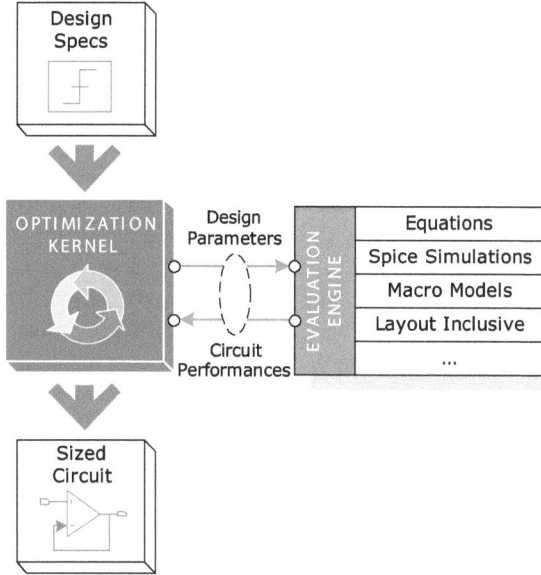

Fig. 3.2 Optimization based sizing

3.1.1 Sizing Task

Before layout, the circuit must be sized. The designer must devise an architecture suitable to implement the system specifications and then all the blocks. From the high-level specifications to the devices sizes all blocks must be properly dimensioned. In industry this task is commonly done manually. The designers start by finding an approximate solution using simplified analytical expressions, and then iteratively, adjust the solution until it meets all specifications.

Another possibility for circuit sizing is the use of circuits' synthesizers. These tools are used to automate the circuit sizing and can be: (1) equation-based, where the methods use analytic design equations to evaluate the circuit performance; (2) numerical-simulation-based, which do not comprise a previous modeling task, but require higher execution times than equation-based due to the verification done using electrical simulations; and (3) numerical-model-based, which use macro models, like neural-networks or support vector machines to evaluate the circuit's performance. A general architecture for automated circuit sizing using optimization based approaches is presented in Fig. 3.2.

One of the critical problems in analog IC design is the process variability, i.e., devices designed to be equal are different after production. To verify if the design is robust, special analysis techniques are employed to ensure that the vast majority of the fabricated circuits will work according to specifications. The most common techniques for analog design centering are Monte Carlo simulation and Corner analysis. Monte Carlo technique executes many simulations by applying random variations to the circuit's and process' parameters, making a stochastic sampling of the behavior of the circuit in real world conditions. Corner analysis is a worst-case approach where the circuit is simulated over extreme combinations of, for instance, process parameters, power supply and temperature values. Sizing ends with some sort of centering analysis to ensure design robustness to technological deviations, reducing the impact of technology gradients and environmental conditions on circuit's performance.

For the analog design flow of Fig. 3.1, the problem of automatic specification translation at circuit level, circuit sizing, is performed by an in-house tool, GENOM-POF [1–4]. This tool is based on an elitist multi-objective evolutionary optimization kernel [5] and uses an industrial grade simulator HSPICE® [6], to evaluate the performance of the design. GENOM-POF targets the design of robust circuits by considering Corner cases during optimization. The inputs are the circuit netlist, testbench, the definition of the optimization variables, design constraints, objectives and the corner cases. The output is a Pareto optimal front of sized circuits, which fulfill all the constraints, and represents the feasible tradeoffs between the different optimization objectives.

3.2 Layout Generation Design Flow

LAYGEN II can support the designer as a standalone tool in the iterative process of layout generation. The layout design flow using the tool as a low-level generator is illustrated in Fig. 3.3. Assuming that the desired technology design kit is available, given the high level floorplan, devices sizes and connectivity between modules the target layout can be automatically generated.

After the first generation, the result can be iteratively re-generated by applying adjustments to the topological relations between cells in the template. In this way, the designer can control the generation from a higher level, and easily introduce new guidelines, until the desired solution is obtained for the current set of devices sizes.

Although in a typical manual IC design process the layout generation task is always performed after the specification translation, often designers have to introduce some changes in the project after the layout is concluded. Those changes may occur by different causes, for example, from a late adjustment to the previous specifications that consequently result in new devices sizes, or even from a sub-block replacement at system level. In the manual process of designing layouts this redesign may lead to partial or complete loss of the previous work, reflecting in the total time necessary to obtain a satisfying solution. In LAYGEN II the information is reused because the guidelines and connectivity are kept in the layout meta-description, making layout adjustment much more efficient.

The same template can be used to retarget a circuit for completely different specifications, being at the discretion of the designer if the old guidelines are still satisfactory for the new design. Although the template may require some adjustments, these changes in the high level floorplan take an almost negligible time when compared to a complete manual re-design. The reuse of old designs is a

3.2 Layout Generation Design Flow

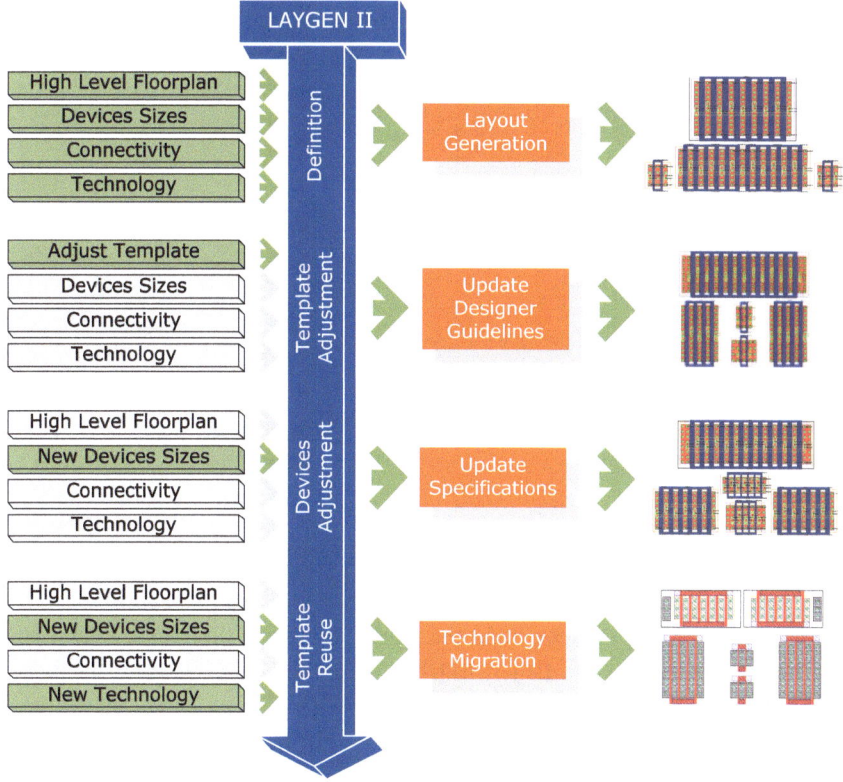

Fig. 3.3 Analog layout design flow using LAYGEN II

common practice of analog designers, both for technology retargeting, using the previous guidelines to generate the circuit for a different technology, as for the reuse of some circuit blocks in a system level design.

In summary, the input parameters can be iteratively changed by the designer, updating high level floorplan guidelines, devices sizes or choosing new technologies to obtain the high quality desired layout. In this design flow, the designer is producing technology and sizing independent layout descriptions that can be used for retargeting, where previous design are (re)used with few or no changes, making the design flow highly reusable and efficient.

3.3 Tool Architecture

The proposed functional architecture is shown in Fig. 3.4 and depicts the principal tasks performed by LAYGEN II. The designer defines a template that provides a high-level circuit description, featuring placement and routing guidelines. The

Fig. 3.4 LAYGEN II general architecture and interfaces

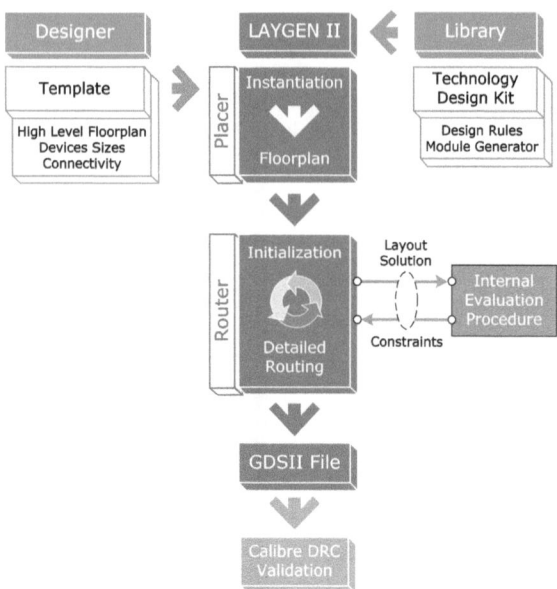

template, the sized components and the target technology design kit, comprising the design rules and a set of parametric module generators, are used by the tool to automatically deal with the exact placing and routing. This automatic generation is performed attending to the specific design rules of the target technology and the device sizes to the target application. The generation proceeds in the traditional way, first placement and then routing, lastly a final DRC validation step is performed. This architecture follows a hybrid solution of correct-by-construction and optimization-based methodology.

The output is a GDSII stream format file, a standard in the microelectronics industry for IC layout data exchange. GDSII is a binary file format composed of several structures that are hierarchically related and a set of elements for each structure. The physical validation of the result is performed in Mentor Graphics' Calibre® [7] DRC tool, a reference in the industrial ICs design.

The Placer encompasses all tasks from the modules instantiation to the procurement of floorplan, following a strict template-based approach. During the instantiation phase, parametric devices are generated using the module generator, custom cells are loaded into the database and sub-templates are generated by the same set of operations on its own modules. When all the device level layouts are available, the placement is performed using the topological guidelines from the template through a set of procedures, automatically merging devices (whenever possible, which will improve the floorplan solution) and ensuring that the floorplan satisfy the technology design rules. The Placer implementation is described in detail in Chap. 4.

The Router uses the obtained floorplan as the starting point, then, the connectivity and the set of symmetry and sensitivity constraints are used to guide an evolutionary optimization kernel. In this process, multiple sequential executions of the optimizations kernel may be used, being that the last execution must result in the detailed and final routing. The routing solution fits in the previously obtained floorplan and ensures that the technology design rules are strictly respected and that it obeys to the set of constraints defined by the designer. LAYGEN II's router follows a highly flexible optimization-based generation approach, which will be described in detail in Chap. 5.

Inside the optimization cycles of router an internal evaluation procedure is used to evaluate each layout solution, mainly because commercial tools' execution times are prohibitive. Still, the parsing of the commercial tool report would mean an extra work when settling a new design kit, deteriorating LAYGEN II's modulation, since a new design kit would require a different parser. The hardcoded design rules are already available for the module generation, so they could be equally used in the internal evaluator. This internal procedure provides LAYGEN II the means required to evaluate if a layout will be successfully validated and thus guide the optimization. After the detailed routing is obtained and the GDSII generated, there is a step of validation with Calibre® DRC tool to assure that the output complies with the design rules.

LAYGEN II has a complete graphical user interface (GUI), so there is no need for the designer to use an external layout design tool to accurately visualize the results. Since GUI doesn't allow editing but only visualization, if the designer intends to manually edit some details on the automatically generated layout, the output GDSII file should be imported in a layout editor, e.g., Virtuoso Layout Editor. The GDSII file provided is an industry standard and is supported by the majority of technology vendors and compatible with nearly all EDA software, commercial or free tools.

In the next sub-sections some details of the implemented GUI and of the supported technology design kits are provided. In addition, a brief introduction to the input parameters and hierarchical capabilities of the high level cell description used by LAYGEN II is also presented.

3.3.1 Graphical User Interface

LAYGEN II's framework and GUI are implemented in Java^{TM} 1.6, which is platform independent programming language. Though efficiency is an important issue, since automatic layout generation using evolutionary computation techniques is under constant development, it is important to keep the modularity and flexibility of the implementation.

As mentioned the tool offers a visualization engine, in Fig. 3.5 is displayed a screenshot of the GUI. This GUI provides a simple and fast way for the designer to check the evolution of the automatic generation. This is particularly important in the development of the tool, since the impact of introducing new features is easily evaluated in all stages of the generation. Black-box generators are not suited for

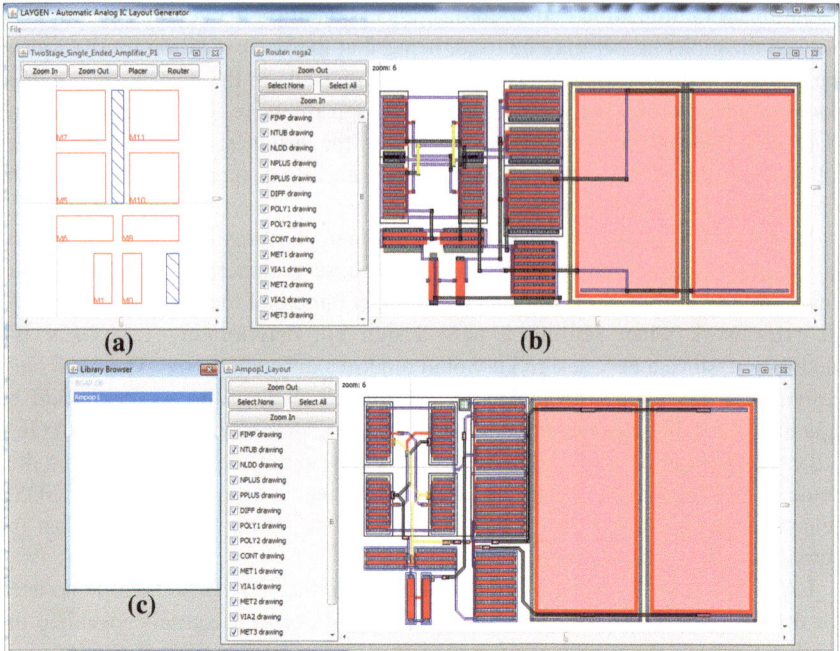

Fig. 3.5 LAYGEN II GUI: **a** Template viewer, **b** Layout viewer and **c** Library browser

multiple reporting facilities, and given the graphical nature of layout generation justifies the use of an own interface.

The template used in LAYGEN II is quite suitable for graphical display. It is provided a graphical view of the topological relations between cells of the designers' defined templates, this way the errors in the template definition are easy to identify. The template viewer depicted in Fig. 3.5a provides scroll and zoom functionalities; launchers for the placer and router are also available.

The layout viewer uses the display settings associated to the layout's technology design kit to define the graphical properties, such as color, drawing pattern and z-axis value for each layer. In Fig. 3.5b shows an automatically generated layout. The layout viewer also implements scroll and zoom functionalities, it provides automatic zoom adjustment to the displaying window when is maximized or resized, and it is also possible to select or unselect a layer for display. The internal evaluation procedure of router automatically places markers in the layout reporting the errors detected, this is essential for the designer to identify the errors, but also for the developer debug the tool.

The GUI can also be used as GDSII viewer. The files can be imported and a library browser is provided to easily explore the current layout hierarchy. In Fig. 3.5c an imported hand-made design and the associated library browser are presented.

3.3.2 Technology Design Kit

The modules during placement are either, generated by the target technology design kit parametric module generator, sub-templates that will be generated during the instantiation phase of the main generation or custom hand-made layouts imported from GDSII files. Special handling is required when designing custom cells to be used in LAYGEN II, since it is necessary to identify the terminals and pins of the cell in a manner consistent with the tool. Nevertheless, the use of an industry standard allows the use of previous designs as modules whenever suitable, and there is no limit on each cell complexity, a cell can be as simple as transistors and as complex as amplifiers, because they are used as macro-cells by the LAYGEN II.

In order to define an internal technology design kit it is required: a layer structure with the enumeration of the layers and vias available; a layer map, containing the GSDII number and type of each layer; the GUI display settings, which describes the colors and patterns used by the layout viewer; the design rules that the technology must comply; and the parametric module generator. Currently, two technology design kits are available in the tool, namely, a design kit for a 350 nm process, and also, for a 130 nm process, which is used over this book to generate the majority of examples and test cases.

The parametric module generator is usually used to instantiate the basic available devices. These modules are described in terms of technology design rules in order to facilitate porting between technologies. Even though there might be large differences between technologies, which are more remarkable when changing to a different factory or distant technology nodes, much of the effort spent when defining the parametric modules can be reused when settling a new design kit. This reuse may imply only some hours of work, while defining a new module generator could require some days to perform. At the time of generation or migration, modules should be validated in Calibre® DRC for a large different number of device sizes, this way it is possible to ensure that all parametric modules comply with the design rules.

Each module can have multiple terminals, and each terminal can have any number of non-overlapping pins. The pins represent the possible points to connect the wires and should be inserted in the physical layer where the corresponding shape is. During the instantiation of the modules, after the rotations (if required), the topological labels are assigned to the pins. These labels provide the absolute location and layer information for the pins composing that terminal in the module being generated.

3.3.3 Hierarchical High Level Cell Description

To reduce the unwanted impact of parasitic and process variations, many knowledge-intensive constraints have been considered in analog design. Device

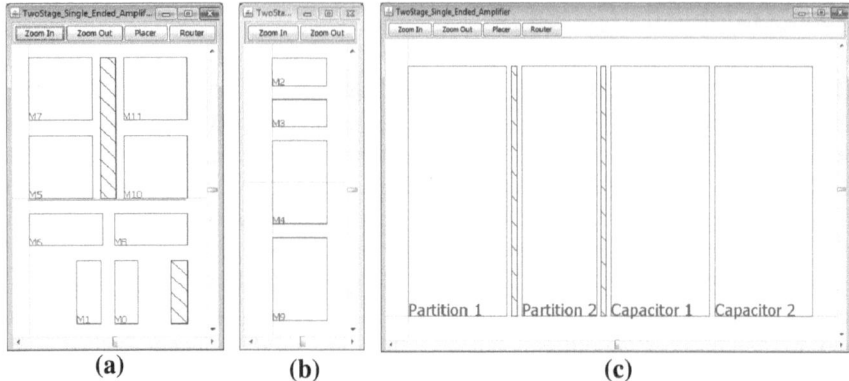

Fig. 3.6 Example of a hierarchical template, partition 1 and 2 are the sub-templates **a** partition 1, **b** partition 2, **c** top partition

matching, symmetry and proximity, current density in interconnects and thermal effects are just some of the factors that analog designers have to consider while planning an analog layout. While designer's expertise is essential in this phase, this knowledge does not require the designer to being aware of the exact details of the technology, e.g., minimum distances or enclosures allowed between layers, etc.

Designer's expertise is caught into the technology independent template and used to guide the automatic layout generation. These guidelines will provide to LAYGEN II the necessary information to increase the layout quality and to achieve a satisfying solution for the designer. The template must allow the tool to generate the target layout in a technology and specification independent way. An extensible markup language (XML) [8] description is used to define the template.

The information used for placement is the devices sizes and the high level floorplan contained in the template, namely the topological relations between cells and a set of symmetry and matching requirements. The topological relations are used to generate a B*-Tree layout representation [9], which the extraction and packing procedure is described together with placement in Chap. 4. For routing the designer provides the connectivity between devices and set of symmetry and sensitivity constraints, if desired. This information is used to automatically generate a routing solution that complies with the design rules, in a process detailed in Chap. 5.

Templates can also be used as modules in a hierarchically manner, allowing the designer to use templates for simpler cells in the definition of more complex ones, splitting the complexity of the space search into different executions of the optimization kernels. The hierarchy is explored as a bottom-up approach, where the physical representation of the lowest levels must be available or generated before proceeding to the automatic generation of the higher ones in the hierarchy. It is also possible to perform routing in the sub-templates from a top template, considering the hierarchy in the description of the connectivity. In Fig. 3.6 it is provided an example of a hierarchical template definition, each of the blocks represented may have any number of sub-templates, and so on.

Despite the advantages, the placement constraints enforced in the template, which are defined by the designer, may inhibit the performance of the target layout. For wide specification changes the topological relations between cells may not lead to a suitable layout, as some new arrangement of modules may be required. For its part, the connectivity provided for routing does not have to be modified due to any changes performed in high level floorplan.

3.4 Conclusions

Analog IC layout design is complex and yields long development times. In this chapter the LAYGEN II tool architecture was presented, that aims to reduce the design time by the introduction of a design methodology that is inherently technology and specification independent. The template description employed creates an abstraction level between physical representation and designer's knowledge. The designer has a way to control the process at a higher level, leaving LAYGEN II responsible for dealing with the exact placement and routing, while attending the specific design rules of the target technology and the device sizes specific to the target application. In addition, the support for hierarchically defined templates allows the designer to define complex cells at expense of simpler ones.

LAYGEN II presents itself as a tool to assist the designer in the generation of layouts and the solution obtained from the intelligent pruning of the design space can be used as a first cut solution. The methodology focuses on the efficiency of retargeting operations, introducing new guidelines, update specifications or migration of designs to different technologies are now tasks easier to perform in a project supported by LAYGEN II. The detailed description of the techniques used for computing placement and routing is found in the next chapters.

References

1. M. Barros, J. Guilherme, N. Horta, GA-SVM feasibility model and optimization kernel applied to analog IC design automation, in *ACM Great Lakes symposium on VLSI*, pp. 469–472, Mar 2007
2. M. Barros, J. Guilherme, N. Horta, *Analog Circuits and Systems Optimization Based on Evolutionary Computation Techniques, Studies in Computational Intelligence*, vol. 294, (Springer, Berlin, 2010)
3. M. Barros, J. Guilherme, N. Horta, Analog circuits optimization based on evolutionary computation techniques. Integr. VLSI J. **43**(1), 136–155 (2009)
4. N. Lourenço, N. Horta, GENOM-POF: Multi-Objective Evolutionary Synthesis of Analog ICs with Corners Validation, in *International Conference on Genetic and Evolutionary Computation Conference (GECCO)*, Jul 2012
5. K. Deb, A. Pratap, S. Agarwal, T. Meyarivan, A fast and elitist multiobjective genetic algorithm: NSGA-II. IEEE Trans. Evol. Comput. **6**(2), 182–197 (2002)
6. Synopsis, http://www.synopsys.com

7. Mentor Graphics, http://www.mentor.com
8. XML, http://www.xml.com/
9. Y.-C. Chang, Y.-W. Chang, G.-M.Wu, S.-W.Wu, B*-trees: A new representation for nonslicing floorplans, in *Proc. 37th ACM/IEEE Design Automation Conference (DAC)*, pp. 458–463, 2000

Chapter 4
Placer

Abstract This chapter presents the methods used by the placer to process and place the modules in the floorplan while following the designer guidelines embedded in the template. The general architecture of the placer is addressed followed by the description of the high level guidelines present in the template. Finally, the detailed generation procedure for the floorplan, depicting each task implemented in LAYGEN II's template-based placer is presented.

Keywords Automatic analog IC layout generation · Abutment · B^*-tree layout representation · Guard ring · Placement · Template-based generation

4.1 Placer Architecture

The proposed synthesis architecture is shown in Fig. 4.1, which depicts the tasks performed by the template-based placer: Instantiation, Pre-Processing and Post-Processing. All these stages terminate with a packing operation, where the topological relations present in the template are mapped to the non-slicing B*-tree layout representation [1], on which the $O(n.log(n))$ packing algorithm presented in [2] is used to obtain a compact placement.

The usage of a template that incorporates expert knowledge about the high level floorplan allows LAYGEN II to present the designer with a meaningful and designer oriented solution. The placer acts as a macro-cell placer without overlap, whose function is to increase the solution quality while obeying to the template guidelines. Even though the Placer does not create any nets or wires, the connectivity is necessary to increase the quality of the obtained placement.

In the previous LAYGEN implementation [3], an optimization kernel (usually simulated annealing (SA) [4]), was used to find the optimal combination of modules that minimize the effective area occupied, or if desired, restricting the

Fig. 4.1 Template-based placer architecture

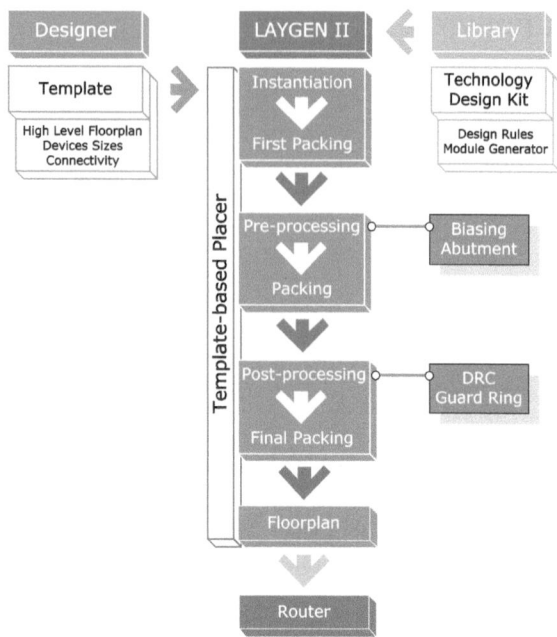

obtained placement to a certain aspect ratio. The placer explored the alternative placements by selecting one of the alternatives from a set different layout representations for each device and packing the corresponding target layout. These alternative modules were for example, a range of possible number of fingers for the same transistor or different aspect ratios for a capacitor. LAYGEN II still support the SA optimization kernel, but the truth is that designers want the devices to match exactly what was simulated, and in practical design they do not want alternatives for the devices. For example, the transistor models for electrical simulators consider the number of fingers so they are weighted during sizing task, and changing then only for layout will introduce untested effects, therefore, given this feedback from the designers, the functionality is no longer used and its implementation is not described here.

4.2 Template

The template information used for placement is the type and relative placement of the cells and the symmetry and matching requirements. The connectivity between cells is also used by the placer. In the template, the floorplan of the cells is described by a box shape, the sizes of this box have no meaning, only the relative positioning between cells (boxes) is of concern, and the topological constraints are inferred from the template boxes' placement directly. Symmetry is specified by two properties, symmetry group that defines a set of cells that share the same symmetry axe; and symmetric cell, the cell that is placed symmetrically in relation to the symmetry axe, while matching defines a set of cells that the designer deemed suited to be matched.

Minimum spacing technology design rules are obviously ensured between modules, however, the designer may want to force some gaps between cells for routing, this is possible by using the Routing Channel construct. The Routing Channel is treated during placement as any other cell, except that its layout representation is an empty box. This allows the designer to specify areas intended for routing, which eliminates the need to consider wiring locations during the packing operation.

With all the information present in the template, the initial coding can be time consuming, but is performed only once for each circuit and all subsequent changes in the designer guidelines result only from small adjustments. Figure 4.2a presents a template description for the layout generation of the differential amplifier with a current mirror load [5] of Fig. 4.2b. The template view for the provided topological relations between cells is shown in Fig. 4.2c. This example will be used through this chapter as a demonstration example for the operations performed by the placer.

A previous sizing task was performed by GENOM-POF [6–9] whose optimization objectives were minimizing the estimated area and power, and maximizing the gain. The first point of the obtained Pareto optimal front, i.e., the sizing solution which minimizes the area was selected to be used in this chapter. The device sizes of the presented solutions are presented in Table 4.1 and all the layouts presented over this chapter were generated for the 0.13 μm UMC (United Microelectronics Corporation Group) design process, which allows a minimum of 120 nm for the channel lengths of the transistors.

4.3 Template-Based Generation Procedure

Before going to the main LAYGEN II's placer tasks Instantiation, Pre-Processing and Post-Processing, detail is first provided about the packing operation used after each of those tasks. LAYGEN II represents the topological constraints used to pack the layout in a B*-tree layout representation. The binary tree representation imposes vertical and horizontal positioning constraints: (a) each device in the left sub-tree is above its parent device and (b) if the y projections of the two devices are overlapping, the device of the node visited first in a pre-order traversal of the tree (visit any node before its left and right sub-trees) is to the left of the device whose node is visited second [1].

The template provided by the designer is processed to extract the binary tree that encodes the specified constraints. In the extraction procedure, first, all cells are placed in a list, the bottom-left cell is added to the root of the B*-Tree and removed from the list, this procedure is repeated until the list is empty. When the

(a)
```
<CellList>
    <Cell name="M3" symGroupId="1" symCellId="1">
        <Box x="0000" y="3000" w="2500" h="2000" />
        <ParametricCellView device="MOSFET" type="P" bulk="vdd"
            width="17.2" length="0.350" m="6" angle="0" />
        <Match cell="M4"/></Cell>

    <Cell name="M4" symGroupId="1" symCellId="1">
        <Box x="3000" y="3000" w="2500" h="2000" />
        <ParametricCellView device="MOSFET" type="P" bulk="vdd"
            width="17.2" length="0.350" m="6" angle="0" />
        <Match cell="M3"/></Cell>

    <Cell name="M1" symGroupId="1" symCellId="2">
        <Box x="0000" y="0000" w="1500" h="2500" />
        <ParametricCellView device="MOSFET" type="N" bulk="none"
            width="17.6" length="0.270" m="4" angle="0" />
        <Match cell="M2"/></Cell>

    <Cell name="M2" symGroupId="1" symCellId="2">
        <Box x="4000" y="0000" w="1500" h="2500" />
        <ParametricCellView device="MOSFET" type="N" bulk="none"
            width="17.6" length="0.270" m="4" angle="0" />
        <Match cell="M1"/></Cell>

    <Cell name="Mbias3" symGroupId="1" symCellId="-1">
        <Box x="2000" y="1500" w="1500" h="1000" />
        <ParametricCellView device="MOSFET" type="N" bulk="none"
            width="3.4" length="0.120" m="2" angle="0" />
        <Match cell="Mbias4"/></Cell>

    <Cell name="Mbias4" symGroupId="1" symCellId="-1">
        <Box x="2000" y="0000" w="1500" h="1000" />
        <ParametricCellView device="MOSFET" type="N" bulk="gnd"
            width="3.4" length="0.120" m="2" angle="0" />
        <Match cell="Mbias3"/></Cell>
</CellList>
```

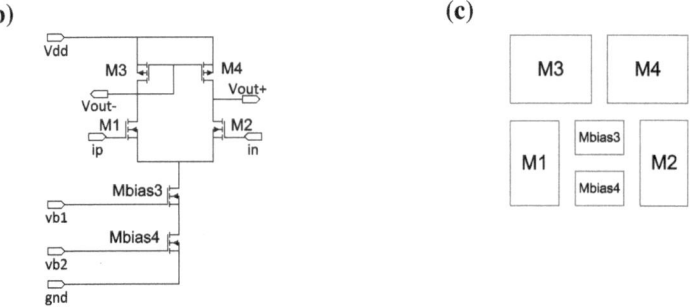

Fig. 4.2 Template example **a** Template description. **b** Diff-amp with a current mirror load. **c** Template graphical view

y-projections of the cell being inserted and the cell of the current node are overlapping, if the cell is to the right of the node's cell, the cell is added to the right sub-tree, if the cell is to the left of the node's cell, the cell replaces the node's cell,

4.3 Template-Based Generation Procedure

Table 4.1 Measures and devices sizes attained during sizing task, using GENOM-POF for the 0.13 μm UMC design process

Measures	Devices	Width (μm)	Length (nm)
Power = 1.18 mW	M1, M2	17.6	270
DC gain = 30.47 dB	M3, M4	17.2	350
Estimated Area = 22.37 μm^2	Mbias3, Mbias4	3.4	120

and the node's cell and the cells in its sub-trees are added to the current sub-tree. When the cell is above the node's cell, there are two scenarios. The cell is above with some x-projection overlapping, in this case the cell is placed in the left sub-tree. Alternatively, the cell is to the right of the current node, in this case there are two possible B*-Tree encoding. The tree is copied, and the cell is placed in the left sub-tree in one copy and in the right sub-tree in the other. Whenever the guidelines are changed, a new B*-tree must be extracted before packing is done.

To generate the target placement, the B*-Tree is packed using the sizes of the selected modules. The packing is performed in two steps, the y-coordinate of the cells are calculated in one pre-order transversal of the B-Tree, each cell positioning is set by knowing the position of its left parent. Then, the x-coordinate of each cell is computed using the Red–Black interval tree algorithm [2]. With the y-coordinates already assigned, the modules are placed in the smallest available x-coordinate that do not yield any overlap. Since multiple different B*-trees may be extracted, each one must be packed for the current devices' sizes and the one that represents the smallest area is selected. The next sections are devoted to the explanation of LAYGEN II's placer tasks, starting with the Instantiation task.

4.3.1 Instantiation

In the instantiation step, all modules contained in the template are substituted by their physical layout representation. The modules can be internal procedural generators from the target technology design kit, sub-templates that will be generated during the instantiation phase of the main generation or custom hand-made layouts imported in GDSII format. All the devices from template example in Fig. 4.2 are generated from the parametric module generator of the current 0.13 μm design kit. The three different modules instantiated are shown in Fig. 4.3.

The instantiation task terminates by executing the packing of the template using these modules. The layout obtained after the first packing is presented on Fig. 4.4.

4.3.2 Pre-Processing

The pre-processing tasks are performed over the layout obtained after first packing. They treat problems related to the substitution of the devices by different ones and consequently handling the alterations in the nets. Pre-processing intends to

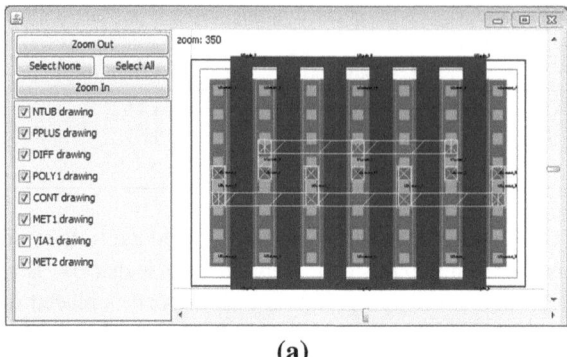

Fig. 4.3 Instantiation of devices from the template example. **a** Transistor M3/M4 (W = 17.2 µm, L = 350 nm). **b** Transistor M1/M2 (W = 17.6 µm, L = 270 nm). **c** Transistor Mbias3/Mbias4 (W = 3.4 µm, L = 120 nm)

automatically implement common techniques employed by the designers to increase the quality of the generated placement. These techniques are applied only for modules generated from the parametric module generator of the technology design kit, and are not applicable to imported GDSII layouts or sub-templates.

4.3.2.1 Biasing

In a layout design, it must be ensured that the biasing is as close as possible to the active devices. The noisy signals affecting the substrate or the well are sunk by the biasing and scarcely affect the circuit itself. Typically, any unused silicon space should be used for biasing purposes [10]. From the viewpoint of manual design, it

4.3 Template-Based Generation Procedure

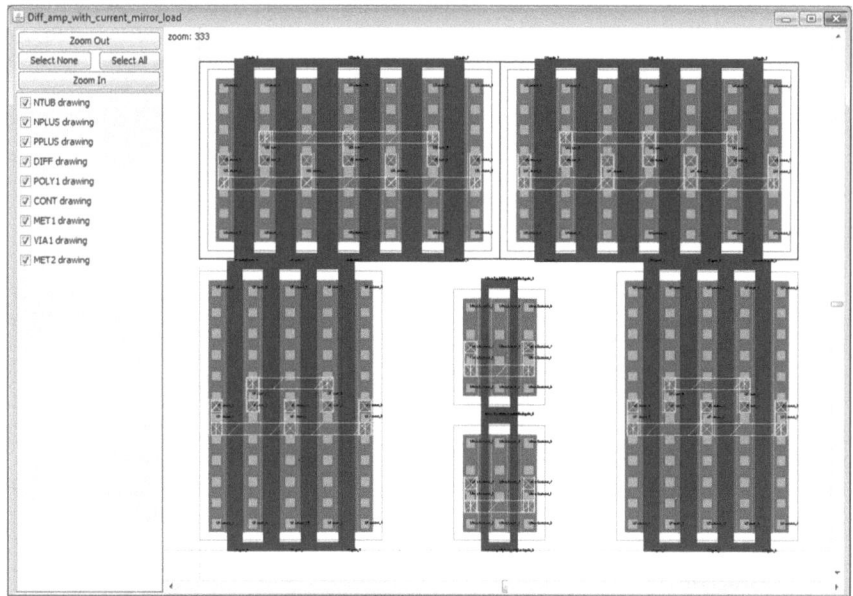

Fig. 4.4 Layout after Instantiation packing

is possible to visualize where the biasing considerations will be placed, often using guard rings for that purpose.

For the automatic layout generation, it is necessary to ensure that there is enough space to add the biasing connections. If the biasing considerations are done after the modules' placement, in some cases there might not be enough space to place them close to the active devices, besides the reduction in performance, this may even violate technology design rules.

The solution used by LAYGEN II to address this problem is to include biasing in the module generation. Particularly, in the actual module generator from the 0.13 μm process it is possible to perform biasing with PMOS transistors, NMOS transistors and guard rings. In Fig. 4.5 some examples of PMOS and NMOS transistors are presented featuring well and substrate contacts, accordingly. For the transistors of Fig. 4.5a and b the well/substrate contacts are merged with the source active area of the transistors.

The Placer starts by using the bulk information provided in the template and analyzes the connectivity to verify if the net assigned to the bulk is the same net assigned to the source. The connectivity from all template hierarchy must be considered, since this connection may exist in higher level templates. The modules that have the bulk at equal potential of source allow the use of topologies of Fig. 4.5a and b. These modules are essential for routing because there is no need to add bulk terminals in the device, it is only necessary to keep the connection to the device source. All modules that suit these requirements are replaced, in the template, by a module with biasing. Since numerous transistors may be replaced in this

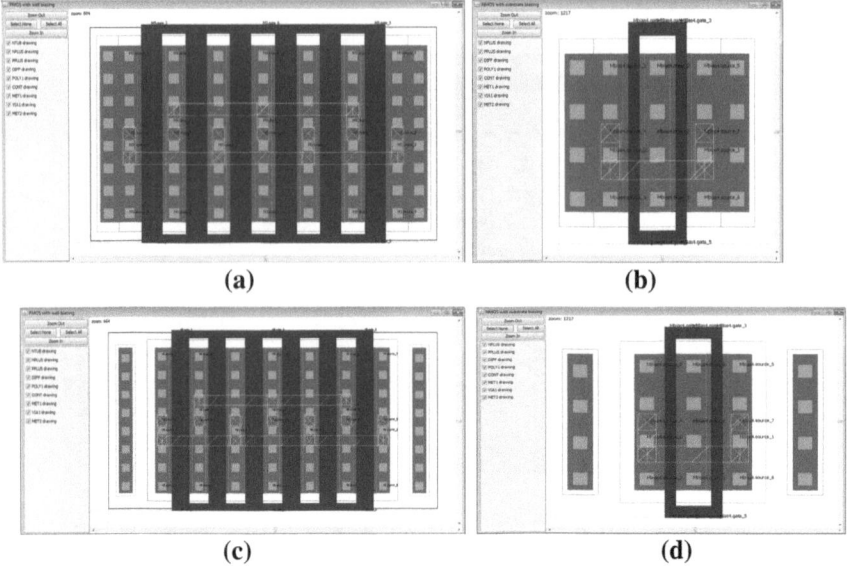

Fig. 4.5 Transistors with well/substrate contacts: (**a,b**) merged and (**c,d**) separate. **a** PMOS with well biasing. **b** NMOS with substrate biasing. **c** PMOS with well biasing. **d** NMOS with substrate biasing

process, the designer can set the bulk information as *none* in the template description for any module, and no device replacement to including biasing is performed for those modules.

In case the net assigned to bulk isn't the same net assigned to source terminal, the module is replaced by one of the topologies of Fig. 4.5c and d. Those modules possess two additional bulk terminals, so it is necessary to process the connectivity for all template hierarchy and add the connection between the bulk terminals and the correct net. If the designer wants to override these automatic considerations, the connection to the bulk terminals can be defined in the template.

4.3.2.2 Abutment

Matched devices appear separate from one another in schematics but they can be combined in the layout. Abutment creates an overlapped connection between two cells without introducing design rule violations or connectivity errors, which not only saves space and reduces the wiring length, but in some cases also improves performance by decreasing parasitics. It is the designer's task to weigh the benefits of the abutment against the possibility of introducing unexpected interactions between the two merged devices [11].

The B*-Tree, although very efficient, is not suited for merging devices. One possible implementation would had been to deceive the packing algorithm of the B*Tree, by providing bounding boxes' sizes that are smaller than the actual layout

4.3 Template-Based Generation Procedure

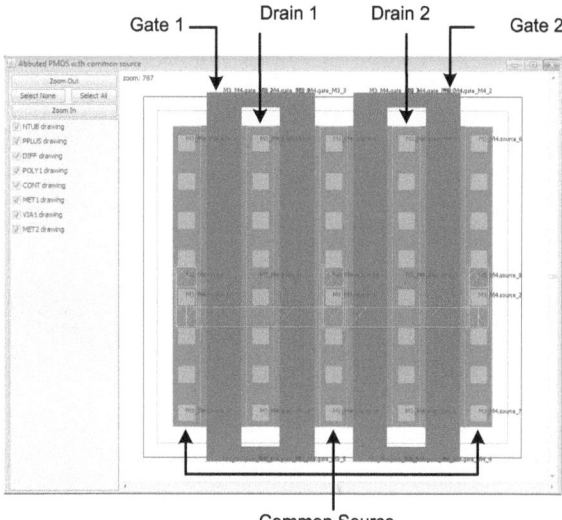

Fig. 4.6 Two PMOS transistors with the source merged

for a cell. However, this may create undesired overlaps, contradicting the macro-cell placer without overlap nature of the approach, forcing design rule validation to ensure all modules were correctly placed.

Another approach is to extend the parametric module generator to support merging devices instead of using the standard basic cells, the tool automatically identifies in a layout the cells suited to be merged and replaces them for a specific parametric module.

If two matched cells are placed together (without any cells between) they are suited to be merged. The placer must analyze the connectivity provided and verify if the two transistors are connected in the same net, both by source or by the drain. As above, the connectivity from all template hierarchy should also be considered. If the two share the same source, a common-source topology is adopted, otherwise, a common-drain topology is used. The gate connectivity is simultaneously verified, if exists, the module is generated already featuring connection between the two transistor gates, this feature is important to automatically create pins between the two sides of the abutted transistor, preserving symmetry. If no terminals are shared, the cells are kept the same, and no changes are performed.

If the two transistors are to be abutted, the two modules in the template are replaced by a single module, as depicted in Fig. 4.6. The new module is formed by the two previous modules bounding boxes, as well as their names. This process is repeated for all pairs to be merged. At this stage it is necessary to update the connectivity and constraints for all template hierarchy to ensure coherence after the replacement of modules.

For the example of Fig. 4.2 the template obtained after applying this processing is presented in Fig. 4.7. Although transistor M1 and M2 are matched and possess the sources connected to the same net, the transistors between prevent them from being merged. Even for a small circuit like the differential amplifier used, with just

Fig. 4.7 Template after abutment processing

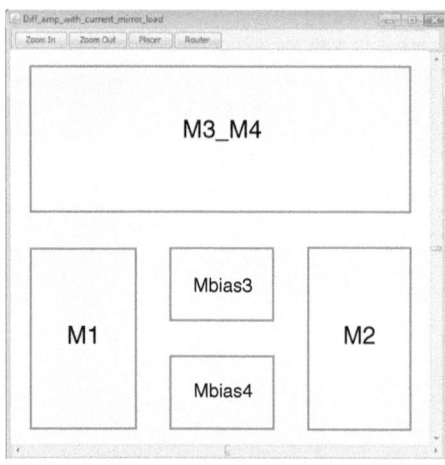

the abutment of transistors M3 and M4 it was possible to remove two wires from connectivity, namely, the connections between the two transistors sources and the two gates.

Besides saving space, abutment also reduces the length of the interconnect wiring, since in this process some wires disappear and thereby obtaining a simpler final layout, for a circuit with large dimensions this reduction in the number of wires will significantly increase the performance of the optimization kernel of the router. This process only allows merging physically identical transistors so that symmetry is kept. The feasibility and advantages of merging devices with different number of fingers or even different widths, should be studied in a future implementation. The same analogy of this section may be utilized to interdigitized or common centroid transistors, if the proper cells are available in the module generator of the target technology design kit.

Since in this pre-processing the template cells are changed, it is necessary to perform a new packing operation. Figure 4.8 presents the obtained layout for the current example after the pre-processing tasks.

4.3.3 Post-Processing

The post-processing tasks are performed over the layout obtained after pre-processing and consequently second packing. In this phase it is not performed module substitutions neither changes to the nets, but rather the current layout is processed in way to verify technology design rules, as depicted next. Then, the guard rings are added if desired by the designer.

4.3 Template-Based Generation Procedure 51

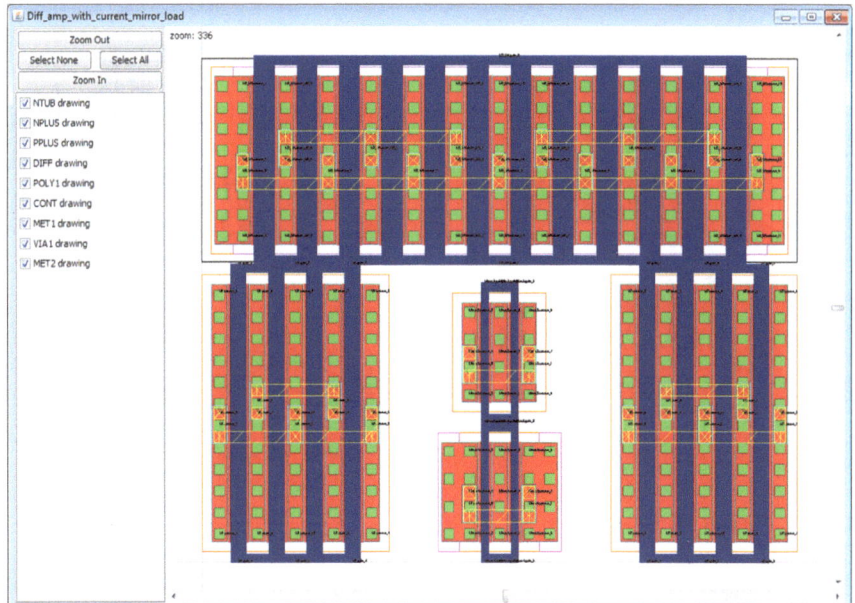

Fig. 4.8 Layout after the packing from pre-processing (GUI)

4.3.3.1 Minimum Distances

As it is possible to observe in all layouts presented until this section, the floorplan modules are all placed together and the minimum distances required by target technology are not ensured. These minimum distances rely on the type, biasing properties and effective location of the devices, so they can only be set when all the modules are known.

This process assumes that cells in the floorplan will not suffer further changes. Each cell in the floorplan verifies the distances between her and all the cells placed above and at right. If the minimum distance is not ensured, the cell's bounding is expanded to the value that fulfills the technology design. This way, the cells are placed exactly at the minimum distances allowed by target technology and no space is occupied beyond the necessary. Since the floorplan provided for post-processing present all modules placed together, some special cases are taken into account without changing the current cells bounding boxes, e.g., wells at the same potential are not separated.

After all the cells are verified, the floorplan is again packed for the new bounding boxes. Each cell is placed at the bottom left of their own expanded bounding box, thus only the need of verifying the cells conflicting at top and right. This task assumes that all modules instantiated were previously validated by Calibre® [12] DRC, so there is no need to verify internally in the modules the minimum distances, enclosures or extensions between layers.

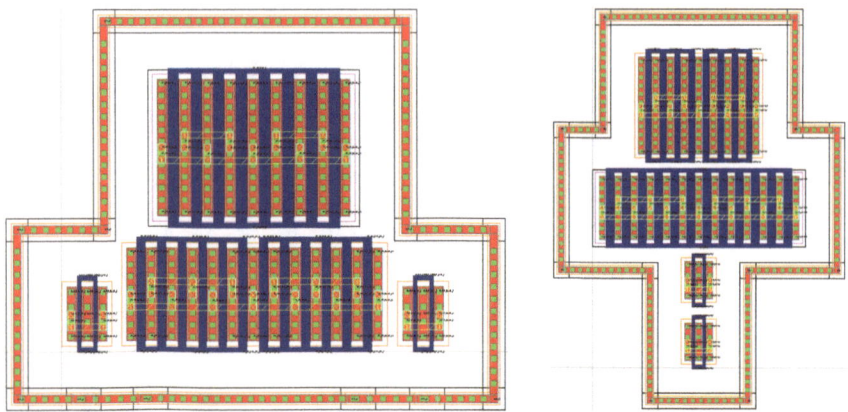

Fig. 4.9 Examples of the automatic guard ring adjustment to the obtained floorplan

4.3.3.2 Guard Ring

Of all the types of failures that plague integrated circuits, one of the more frustrating is definitely latchup. Devices that operate properly in one circuit, may latchup when are inserted into another for no apparent reasons. This problem is aggravated since simulation and most forms of testing rarely uncovers latchup problems. The smaller dimensions of modern complementary metal oxide semiconductor (CMOS) processes made them more prone to latchup. Although not totally solve the problems, the use of guard ring try to enhance immunity to this possible circuit failure and should be used by designers as suited [11].

The possibility of generating guard rings with different shapes, when supported, is a feature commonly used by the designers in the layout edition tools. In LAYGEN II is equally possible to automatically generate guard rings with different geometries. In order to define which guard ring geometry must be drawn, the solution found was to take all the four edges of the bounding boxes of all devices in the floorplan, already considering the minimal distances between each cell, and applied an algorithm of convex hull.

In the current planar case, the convex hull for a set of finite points is the minimal convex polygon containing all the points. The method of computing the convex hull used is the Graham's scan [13], which is a method of computing the convex hull in a plane with a complexity of $O(n \log n)$. For all non 90° segments obtained by connecting two followed points of the polygon, a third point is added to perform only rectangular structures. Some guard ring examples of the versatility provided by the current implementation are shown in Fig. 4.9.

There are two available types of guard rings to be selected by the designer, namely the P plus-based ring or N Plus-based ring. The P plus-based ring is relatively shallow, and his ability to correctly blind the involving circuit is compromised since it can only intercept a fraction of the carriers. The N Plus-based ring although it is placed inside an N-well still has its limitations since most of the

4.3 Template-Based Generation Procedure

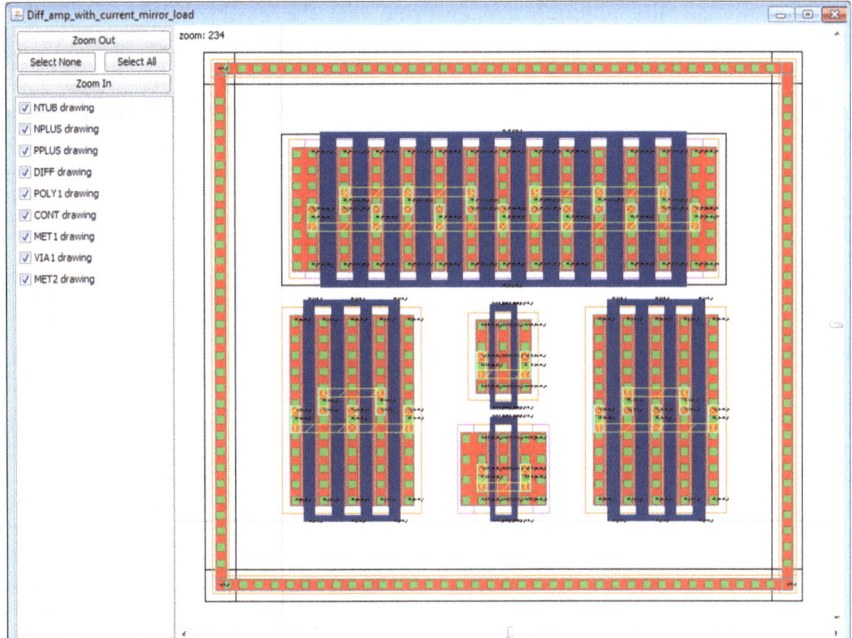

Fig. 4.10 Floorplan obtained

carriers flow down to the substrate instead of laterally to the guard ring. For a sturdy layout should be used a combination of different types of guard rings both to suppress most forms of latchup, and as usually used by designers to correctly bias the circuit.

The actual implementation places the guard ring around the circuit defined by a template. So, if designer only wants one transistor surrounded by a guard ring, it should define a sub-template for that module, taking advantage of the hierarchical capabilities of LAYGEN II. Also, it is important to notice that each guard ring possesses a bulk terminal that should be properly connected during the automatic routing task.

In Fig. 4.10 the layout obtained for the current example after the post-processing tasks is presented, and, consequently, the final floorplan which will be provided for routing.

4.4 Conclusions

In this chapter, the template information required for placement and the adopted generation techniques were introduced. The designer provides the high level floorplan and the tool automatically instantiates and places the devices on the layout, automatically abutting and ensuring that the design rules are fulfilled. Design rules were only mentioned in the post-processing of Placer, as LAYGEN II's placer is a macro-cell placer without overlap. Therefore knowing that the instantiated cells comply with the design rules and assuming that the circuit is properly biased, maintaining the distance between devices is the only operation required to verify the technology design rules. Lastly, the tool automatically adapts guard rings to the floorplan obtained. A simple amplifier was used through the chapter to explain the proceedings.

References

1. Y.-C. Chang, Y.-W. Chang, G.-M.Wu, S.-W.Wu, B*-trees: a new representation for nonslicing floorplans, in *Proceedings of 37th ACM/IEEE Design Automation Conference (DAC)*, pp. 458–463 (2000)
2. F. Balasa, S. C. Maruvada, K. Krishnamoorthy, Using red-black interval trees in device-level analog placement with symmetry constraints, in *Proceedings of the Asian and South Pacific—Design Automation Conference (ASPDAC)*, pp. 777–782 (January 2003)
3. N. Lourenço, M. Vianello, J. Guilherme, N. Horta, LAYGEN—automatic layout generation of analog ICs from hierarchical template descriptions, in *Proceedings of the Conference on Ph.D. Research in Microelectronics and Electronics (PRIME)*, pp. 213–216 (June 2006)
4. B. Suman, P. Kumar, A survey of simulated annealing as a tool for single and multiobjective optimization. J. Oper. Res. Soc. **57**, 1143–1160 (2006)
5. R. Jacob Baker, *CMOS Circuit Design, Layout and Simulation* (IEEE Press, New York, 2005)
6. M. Barros, J. Guilherme, N. Horta, GA-SVM feasibility model and optimization kernel applied to analog IC design automation, in *Proceedings of ACM Great Lakes symposium on VLSI (GLVLSI)*, pp. 469–472 (March 2007)
7. M. Barros, J. Guilherme, N. Horta, *Analog circuits and systems optimization based on evolutionary computation techniques*, Studies in computational intelligence, vol. 294 (Springer, New York 2010)
8. M. Barros, J. Guilherme, N. Horta, Analog circuits optimization based on evolutionary computation techniques. Integration, VLSI J. **43**(1), 136–155 (2009)
9. N. Lourenço, N. Horta, GENOM-POF: multi-objective evolutionary synthesis of analog ICs with Corners validation, in *Proceedings of Genetic and Evolutionary Computation Conference (GECCO)*, (July 2012)
10. F. Maloberti, *Analog Design for CMOS VLSI Systems*, (Kluwer Academic Publishers, Netherlands 2001)
11. A. Hastings, *The Art of Analog Layout*, 2nd edn (Prentice Hall, New Jersey 2005)
12. Mentor Graphics, http://www.mentor.com
13. R.L. Graham, An efficient algorithm for determining the convex hull of a finite planar set. Inform Proc. Lett. **1**, 132–133 (1972)

Chapter 5
Router

Abstract This chapter covers the general description of the Router architecture, followed by the description of the template information necessary for routing, namely, the connectivity and routing constraints. Then the routing generation procedure is explained, depicting each task implemented in LAYGEN II's optimization-based router, with emphasis on the evolutionary computational techniques used. Finally, the internal evaluation procedure used to verify if the routing solutions fulfill all the technology design rules and constraints is detailed.

Keywords Automatic analog IC layout generation · Design rule check · Evolutionary computation · Genetic algorithm · Multi-objective optimization · Noisy signals

5.1 Router Architecture

The Router uses the floorplan solution generated by the Placer, the routing connectivity and a set of symmetry and sensitivity constraints contained in the template as inputs. The solution space is then explored ensuring that technology design rules and designer constraints are respected. Unlike the Placer, the Router must ensure that contacts, vias and wires do not violate any of the design rules. In addition, care must also be taken to avoid unwanted contacts between electrically connected layers, wires shunts or connecting a wire to unwanted shapes of the cells underneath. All of these design rule validations and the huge search space make the routing algorithm extremely complex, and computationally more expensive than placement.

The proposed generation architecture is shown in Fig. 5.1, which depicts the main tasks performed by the optimization-based router. This process is executed in

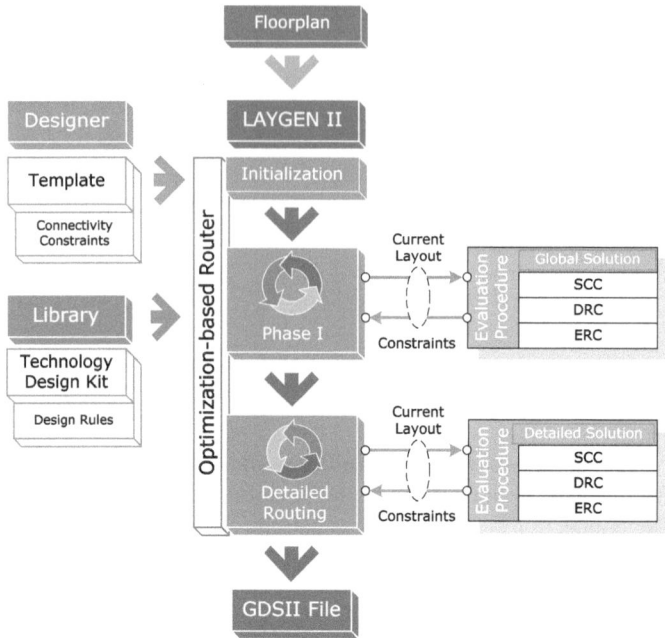

Fig. 5.1 Optimization-based router architecture

two main stages, marked as Phase I and Detailed Routing, preceded by an initialization stage.

LAYGEN II performs a flexible and easy to setup optimization-based routing. The connectivity and constraints although obviously dependent from the circuit, are provided independently from the floorplan attained and the processing tasks done in the Placer. This is, the designer guidelines for placement may change, along with the devices' sizes or even the target technology, but the routing template remains valid. In most cases it is unnecessary to perform any kind of modification to the connectivity or constraints, and only if the topological relations change abruptly some tweaking of the routing template may be required.

In the previous LAYGEN implementation [1] the Router followed a strict template-based approach where the designer provided for each wire in the template the two exact pins to be connected, a geometric shape for the wire which was defined by a set of points, and the preferred conductor layer. All this information made the process of defining the template extremely long. Additionally, the fixed position of the pins and the limited operators, intended to keep to a minimum the deviation from the designer definitions, introduced great limitations on the routing flexibility during generation. Moreover, any change in the topological relations between cells could compromise all the geometries, forcing template re-design. Although suited to abstract designer from technology details, the strict template-based approach did not yield physically achievable layouts for wide specification changes.

5.1 Router Architecture

During the development of the current full-automatic router, the possibility for the designer to totally define the wire geometry was kept. The solution can be generated from any range of template-based wires to automatic generated wires, at the designer's criteria. However, template wires proved to lack flexibility and the automatic wires were strongly conditioned by the enforced geometries, requiring the use of more conductor layers even for a fairly simple routing. The exploration of different placement topologies carried long setup time because the template-defined-wires, while the connectivity for automatic wires does not require any modification. For the reasons outlined, template-based routing is being discontinued and is not described in this document, for further information about this method please refer to [1].

The violations of the design rules are used as constraints during the routing generation, which the evolutionary algorithm must drive to zero. The evaluation is performed by a powerful internal evaluation procedure which possesses three different types of validation, short circuit check (SCC), design rule check (DRC) and electrical rule check (ERC), detailed in Sect. 5.4. Other qualitative measures are used as objectives for the evolutionary algorithm, for example the total wires' length, the number of conductors or contacts used for routing, and other objectives more oriented to the processing of special nets, such as distance between noisy and sensitive.

5.2 Template

Even though the Router follows an approach characteristic of a fully-automatic generator, the connectivity and constraints provided have the template designation, in order to keep uniformity with placer. Each net is divided into a set of wires, each one connecting two and only two contact points, or pins. Internally, each wire is formed by any number of linked segments. For its part, a segment refers to the connection of two different points in the space, whose values in the x axis or in the y axis are the same. The routing connectivity is defined using the same cells labels used for the template of placer, and the terminal labels associated with the modules. The designer only has to concern on defining the set of nets and the respective wires, each one connecting two and only two terminals.

In a cell, the pins of a terminal are identified with a label. The parametric module generators should provide that same label placed over the shapes that are part of the terminal. Even though each terminal may have any number of pins, the router deals automatically with them, the designer only concern is with the terminals. In order to clarify, when referring to the terminals of a cell, e.g., a transistor source, the term terminal is used. When referring the possible contact points of a terminal, e.g., the exact location over the stripe of metal where the connection can be made, the term used is pin. The next paragraphs provide more detail of the elements used to describe the routing template.

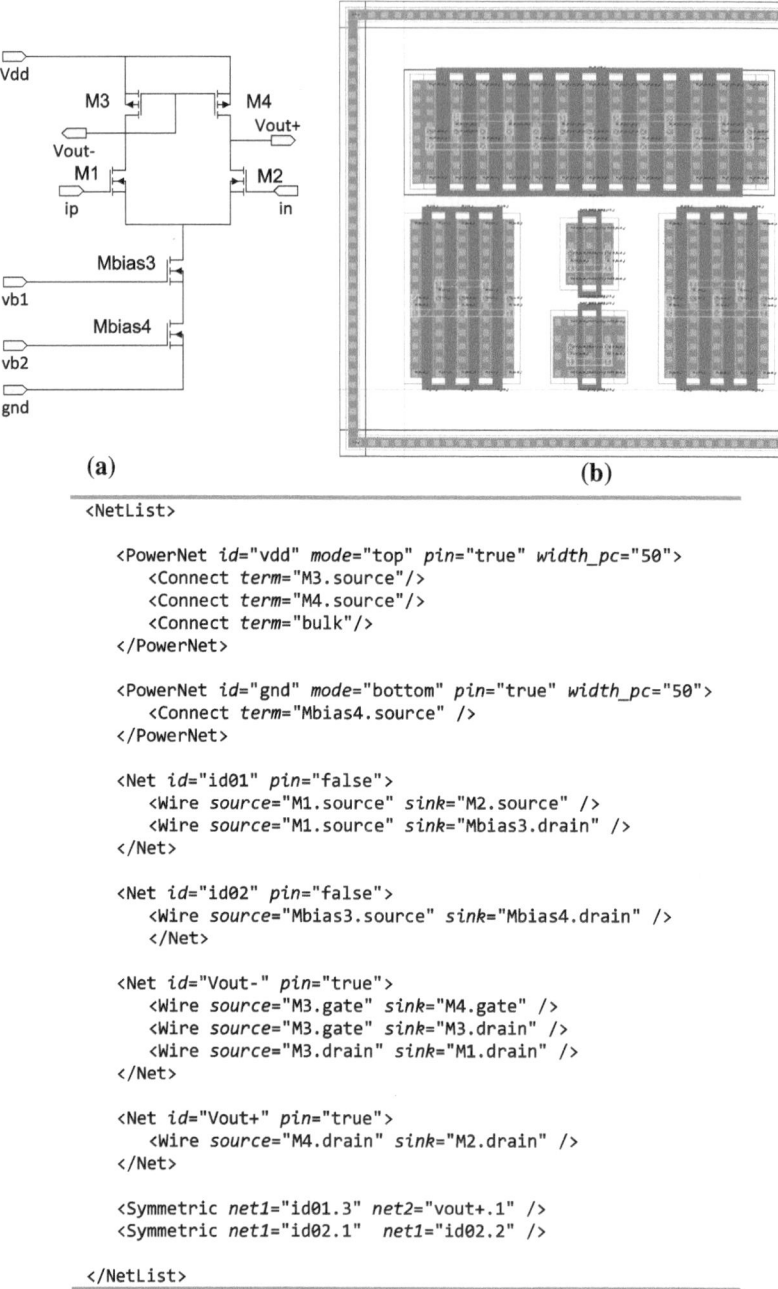

```
<NetList>

    <PowerNet id="vdd" mode="top" pin="true" width_pc="50">
       <Connect term="M3.source"/>
       <Connect term="M4.source"/>
       <Connect term="bulk"/>
    </PowerNet>

    <PowerNet id="gnd" mode="bottom" pin="true" width_pc="50">
       <Connect term="Mbias4.source" />
    </PowerNet>

    <Net id="id01" pin="false">
       <Wire source="M1.source" sink="M2.source" />
       <Wire source="M1.source" sink="Mbias3.drain" />
    </Net>

    <Net id="id02" pin="false">
       <Wire source="Mbias3.source" sink="Mbias4.drain" />
    </Net>

    <Net id="Vout-" pin="true">
       <Wire source="M3.gate" sink="M4.gate" />
       <Wire source="M3.gate" sink="M3.drain" />
       <Wire source="M3.drain" sink="M1.drain" />
    </Net>

    <Net id="Vout+" pin="true">
       <Wire source="M4.drain" sink="M2.drain" />
    </Net>

    <Symmetric net1="id01.3" net2="vout+.1" />
    <Symmetric net1="id02.1"  net1="id02.2" />

</NetList>
```

(c)

Fig. 5.2 Template example. **a** Diff-amp with a current mirror load, **b** Floorplan, **c** Connectivity and constraints description

5.2 Template

```
<PowerNet id="vdd" mode="top" pin="true" width_pc="50">
    <Connect term="M3_M4.source"/>
    <Connect term="bulk"/>
</PowerNet>

<Net id="id01" pin="false">
    <Wire source="M3_M4.gate" sink="M3_M4.drain_M3" />
    <Wire source="M3_M4.drain_M3" sink="M1.drain" />
</Net>

<Net id="out" pin="true">
    <Wire source="M3_M4.drain_M4" sink="M2.drain" />
</Net>

<Symmetric net1="id01.2" net2="out.1" />
```

Fig. 5.3 Template nets changed by the placer's pre-processing task

Figure 5.2c presents a connectivity and constraints description for the routing of the differential amplifier with a current mirror load depicted in Fig. 5.2a and the floorplan of Fig. 5.2b. Here two power nets, *vdd* and *gnd*, were defined that will instantiate a conductor stripe above, on the left, on the right or below the circuit according to the designer's discretion, similar to what it is done in digital layout design. Unlike other wires, the terminals connecting to a power net can be defined sequentially.

Any pair of two wires may be marked as symmetric. Since those two lines are physically identical except one is mirrored, if they do not belong to the same net it is important to ensure that those two lines will not cross each other in the layout, in order not to cause an inevitable short circuit. It is also possible to mark nets as noisy or sensible, to be treated differently from the other nets during the synthesis. The pin flag is activated by the designer if he intends to save the net label in the GDSII file, usually to be used as a terminal for the generated cell.

The lack of electrical measures is compensated by the constraints provided by the designer in the template. It can be defined a certain percentage of extra width to match the current density for a net, resulting that each wire of that net will have a width equal to the minimum width allow by target technology to the current conductor, plus a percentage of that value. For the current template the net *vdd* and *gnd* were defined to be 50 % wider than the minimum width allowed by target technology to the current conductor used. This parameter will affect in the same manner the width of the conductor stripe of the power net.

The connectivity and constraints are provided always taking into account the initial elements from the electrical schematic, and are provided simultaneously with the placement template. As detailed in the previous chapter, the pre-processing tasks perform some changes in the modules and consequently nets. In Fig. 5.3 the changes automatically done by the placer are depicted, these are due to the merging of transistor M3 and M4 sources and gates.

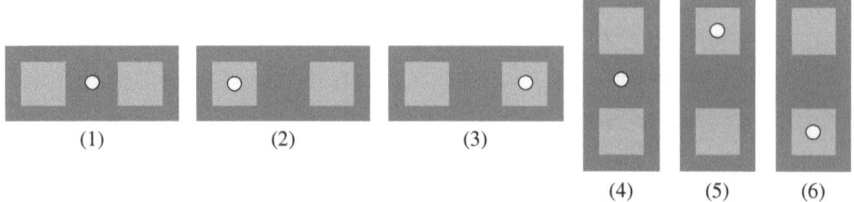

Fig. 5.4 Different arrangements of a contact or via with respect to a connection point

5.3 Optimization-Based Generation Procedure

In the next sub-section before introducing the routing chromosome and the genetic operators, it is necessary to explain how LAYGEN II deals with the transitions between different conductors. Then, some detail is provided for the optimization kernel used, including the chromosome structure, initialization and operators. After, the differences between the two routing optimization phases are depicted, finally, the internal evaluation procedure is explained.

5.3.1 Multiple Contacts

It is a designer's common practice the use of multiple contacts not only on top of the source and drain regions to avoid parasitic transversal drop voltages, but also in any transition between conductors. Usually multiple contacts are placed close to each other at a minimum distance instead of using a single large contact, which may even not be allowed in some technologies. This makes the surface of metal connections smoother, and prevents that the only contact may cause a circuit failure [2]. From the standpoint of automatic generation, the use of multiple contacts is easily achieved for the active regions of the devices, but for small design processes greatly limit the capability of the automatic routing to perform transitions between conductors.

The contacts, vias and connection plates necessary to perform a valid connection between conductors can cause the emergence of more violations of the design rules, hindering the algorithm convergence, so it is necessary to ensure that they are placed in the best possible way. For LAYGEN II's router each transition is performed with a minimum of two contacts or vias. In Fig. 5.4 are depicted the six different arrangements of contacts with respect to a connection point, marked in the figure with the white dot. The contact plate can be placed horizontally or vertically, and centered or shifted to the right or left. The same analogy can be used for more contacts or vias. It is important to notice that the connection point, marked with a dot, may refer to a pin of a device terminal or the point of transition between two different conductors inside a wire structure.

5.3 Optimization-Based Generation Procedure

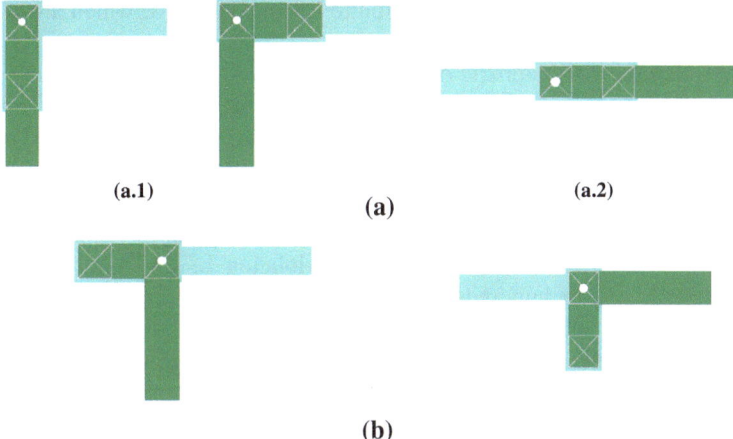

Fig. 5.5 Contact arrangements of a transition between two different conductors. **a** Example of legal contact dispositions, **a.1** Segments with different orientation; **a.2** Segments with the same orientation. **b** Examples of illegal contact dispositions

When optimizing, the router only places the contacts under the wiring connecting. So, for each shape formed with two segments connecting exists just some valid arrangements from all the presented above. When a contact orientation is randomized for a given connection point, this random choice is made within a solution space which contains only the valid orientations, and not all the available, ensuring that a desirable arrangement is enforced.

For each transition between a vertical segment and a horizontal segment, there are two valid possible arrangements, as depicted in Fig. 5.5a.1. For a transition between two segments with the same orientation, Fig. 5.5a.2, there are three different possible arrangements that satisfy the connection, (1) to (3) if segments are horizontal and (4) to (6) if vertical. In Fig. 5.5b some illegal contact dispositions are illustrated. For the connections between a wire and a terminal of a device, the pin label possesses the information that determines if the contact orientation must be horizontal or vertical.

5.3.2 Evolutionary Algorithm

During the initial development of LAYGEN II's router, it was used a classical genetic algorithm (GA) [3] approach where the constraints were modeled simultaneously with the objectives in the fitness function. Given the need to deal with constraints separately from objectives, the kernel was moved to the modified NSGA-II [4], an elitist multi-objective evolutionary algorithm (MOEA). Although it is not being yet explored the full potential of the multi-objective nature of the algorithm, it open up a wide range of future implementations, which are remitted to chapter of conclusions and future work.

The distinctive characteristic of the evolutionary kernel implementation is the chromosome structure. It must be adequate to support complex routing problems and still allow the use of common genetic operators, as detailed in the next section. The constraints are not enforced during the generation of the routing solution, ensuring design correctness for every wire, yielding only feasible solutions. This strategy yields large computation times, and all the infeasible points are treated equally, regardless if there is one violation of the minimum spacing between two wires, or hundreds of design rules violations and unwanted shunts. Instead, the router follows an optimization approach where the population should converge to feasible, as constraints are driven to zero during the optimization procedure. When the physical implementable solutions are obtained, i.e., solutions that do not violate any of the constraints and hence will successfully validate in Calibre® [5] DRC, the optimization objectives are taken into account. These objectives may be, for example, the total wiring length or the total number of contacts (or vias) used, in order to minimize them during the optimization. Each conductor layer has an associated cost pre-defined in the technology design kit, commonly the lowest conductor levels are associated with the lowest costs. So, the wiring length can be computed together with the conductor cost of each segment, in a way to minimize not only the paths but also to favor the use of the lowest conductor levels.

Symmetric routing can be used if the terminals of the wires being routed are symmetric with respect to a symmetry axis. The symmetry is handled at the chromosome level. In a symmetric pair of wires, only one is operated during optimization, the other is generated deterministically from the first.

5.3.2.1 Chromosome

Each element in the population, chromosome, encodes the information of a different routing solution, corresponding each gene to one wire. So, each chromosome has a fixed number of genes equal to the number of wires present in the circuit. In Fig. 5.6a a possible layout solution generated for the connectivity and constrains provided in the example of Fig. 5.2 is presented. In Fig. 5.6b only the wires are represented, and the chromosome formed by nine genes (wires) is presented in Fig. 5.7. For its part, each wire (gene) of the chromosome has the same even number of segments, which facilitates the implementation of the genetic operators. A segment may be defined as horizontal or vertical; if vertical, the one following must be horizontal, and vice versa.

In order to clarify the gene structure, in Table 5.1 the fields contained in a wire structure with four segments are presented. At any time in the optimization, if a segment is defined with a delta length equal to zero obviously does not have physical representation. A layer is always associated with a segment, so a transition between two layers is always related to a transition between two segments.

5.3 Optimization-Based Generation Procedure

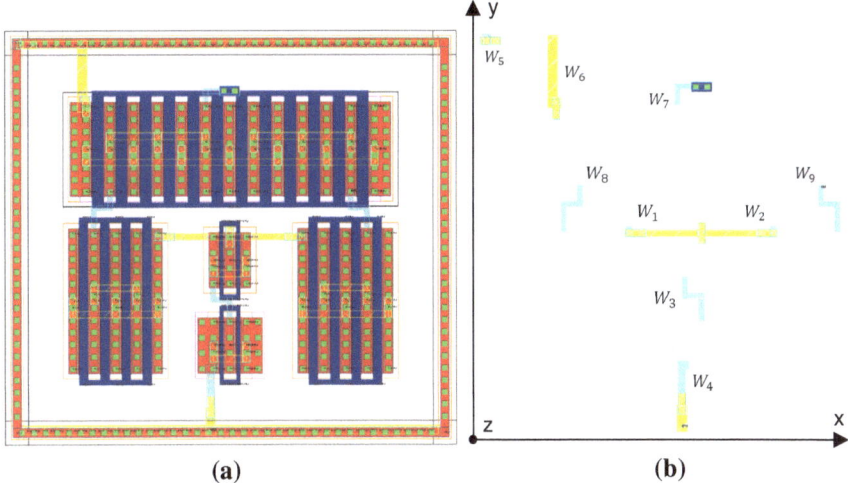

Fig. 5.6 Routing solution. **a** Possible layout solution for the Diff-amp, **b** Physical representation of the wires

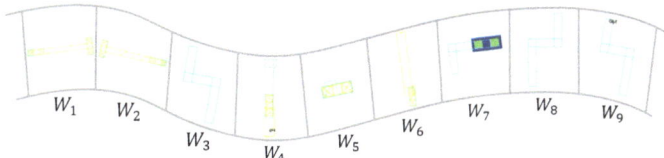

Fig. 5.7 Chromosome used for routing optimization corresponding to the routing solution in Fig. 5.6

Since the current wire structure allows any number of segments with a null length, this permits layer transitions between two segments with the same orientation.

For the examples in this chapter the number of segments in a wire was set to four. While in Chap. 6, given the increasing complexity of the addressed problems and hence the routing solution, each wire was set to be constituted by six segments. In addition to the physical wire structure represented on Table 5.1, the wire also has the information of source and sink terminal that is used to search the best pin location.

5.3.2.2 Initialization

Initialization is the first step of the Router and consists of generating the initial population for the evolutionary kernel of Phase I optimization loop. For each wire of the chromosome only the source and sink terminals are known. The initial pins of the terminals connecting are selected following a greedy approach, where the

Table 5.1 Wire structure composed by four segments

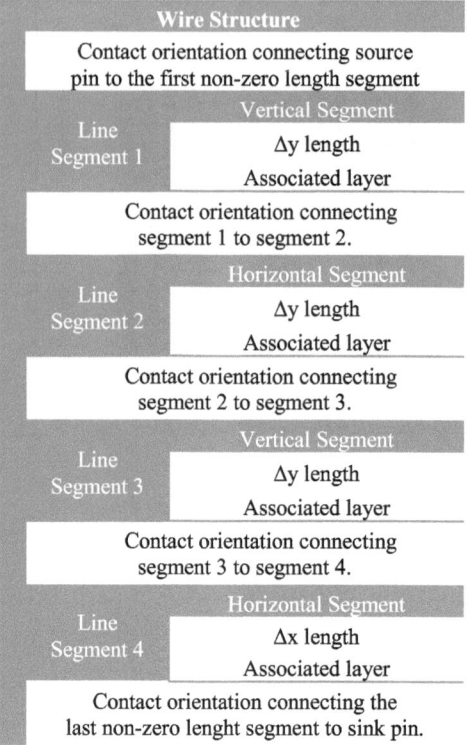

Wire Structure	
Contact orientation connecting source pin to the first non-zero length segment	
Line Segment 1	Vertical Segment
	Δy length
	Associated layer
Contact orientation connecting segment 1 to segment 2.	
Line Segment 2	Horizontal Segment
	Δy length
	Associated layer
Contact orientation connecting segment 2 to segment 3.	
Line Segment 3	Vertical Segment
	Δy length
	Associated layer
Contact orientation connecting segment 3 to segment 4.	
Line Segment 4	Horizontal Segment
	Δx length
	Associated layer
Contact orientation connecting the last non-zero lenght segment to sink pin.	

distance between each pin of the source terminal and each pin of the sink terminal is computed, the pair which represents the minimum distance between the two is selected. If implementable they represent the optimal solution, so are used as the optimization starting point.

Despite the chromosome structure allows fully random generation of the wire segments, it has been found that the use of pre designed shapes, henceforward called heuristics, give a better starting point for the connections between pins. So, each wire is randomly generated through a set of available heuristics; the possible geometries are presented in Fig. 5.8. It is important to notice that in heuristics (c) and (d), the segments inside the wire structure, i.e., those not connected to the source pin neither sink pin, can take any position between terminal pins. Since the only restriction handled in this step is the effective pins location, it only ensures connectivity without validating technology design rules.

Although the wire structure may have a number of line segments higher than four, the use of heuristics with more than four segments has not proved to be necessary. Having the wire geometry set by the heuristic, each wire, and hence all its segments, is initialized with the conductor with the lowest associated cost in the

5.3 Optimization-Based Generation Procedure

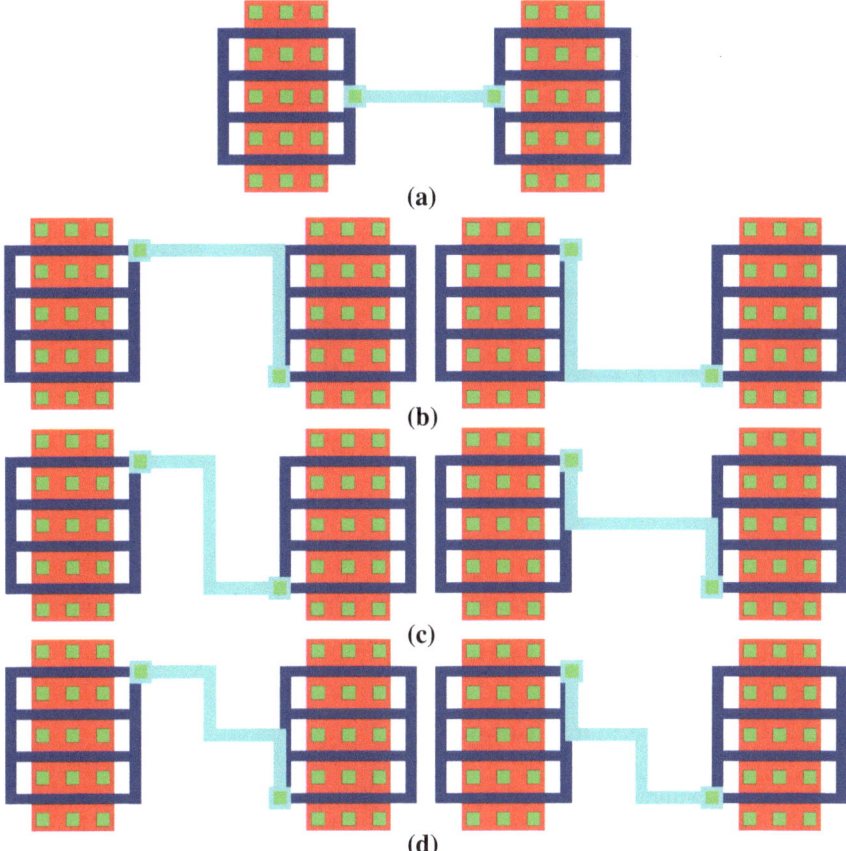

Fig. 5.8 Different heuristics for generating the wires. **a** One segment, **b** Two segments, **c** Three segments, **d** Four segments

technology design kit. Then, each contact orientation of the gene is randomly set within the valid arrangements, ensuring that no connection plate is placed outside the boundaries of the two line segments connecting.

After each wire generation, it is performed a *remove small segments* validation to remove the segments considered too small to be included in the solution. If a segment has a delta smaller than his width, the segment is deleted. Obviously, the removal of this segment will cause the loss of connectivity, so it is necessary to perform a *rescale* step. This processing equally rescales all the deltas from the remainder non-zero segments until the missing value is restored to the wire and connectivity is verified.

In summary, initialization step sets the ideal conditions for the current routing problem. A routing performed with greedy connections using the same lowest cost conductor for all connections would be great to achieve, however it is unlikely to

Table 5.2 Example of crossover, parents (P) and offspring (O)

	Segment 1 Vertical			Segment 2 Horizontal			Segment 3 Vertical			Segment 4 Horizontal			Δ Total	
	Ct	Δ	L	Ct	Δ	L	Ct	Δ	L	Ct	Δ	L	Δx	Δy
P1	5	-156	M1	2	176	M1	5	-156	M1	2	176	M1	4	352	-312
P2	4	0	M2	1	199	M2	5	-312	M2	6	153	M2	6	352	-312
O1	5	-156	M2	1	199	M1	5	-156	M2	2	176	M2	4	375	-312
O2	4	0	M1	2	176	M2	5	-312	M1	6	153	M1	6	329	-312
O1'	5	-156	M2	2	199	M1	5	-156	M2	2	153	M2	4	352	-312
O2'	4	0	M1	2	199	M2	5	-312	M1	6	153	M1	6	352	-312

Ct – Contact orientation (6 possible arrangements, as detailed in Fig. 5.4) ;
Δ - Length;
L – Associated layer: M1 – Metal 1/M2 – Metal 2.

be possible without violating design rules. The removal of the violation rules from the greedy initial population is the optimization problem in question.

5.3.2.3 Genetic Operator: Crossover

Although the evolutionary algorithm follows the typical NSGA-II flow, innovative genetic operators were developed to deal with the specific routing chromosome presented above. At each new generation, each pair of parents is selected by tournament to generate two offspring that present a combination of their information. The parents' contribution refers not only to segments length, but also to the combination of layers and contacts orientation used. A multi-point crossover is used as presented in Table 5.2, for the two random parents (P1 and P2) depicted in Fig. 5.9. Since the offspring represents a random combination of segment lengths from the two parents, denoted as O1 and O2, it is common that the resulting wire fails the connectivity.

After each crossover, three different validations are performed. A *connect* task is done to randomly add (or remove) the missing (or excess) delta x or delta y value to a valid segment, in order to ensure connectivity. Then, a *remove inverted segments* validation is made to assure that doesn't exist two contiguous horizontal or vertical segments with opposite directions. The inverted segment must be removed from the wire structure to guarantee that there are no overlapping segments in a wire. Finally, a *verify contacts orientation* task is performed to ensure that the contact orientations are valid for the new combination of deltas. If not, a new contact orientation within the valid possible arrangements ensuring that no connection plate is placed outside the boundaries of the two line segments it is

5.3 Optimization-Based Generation Procedure

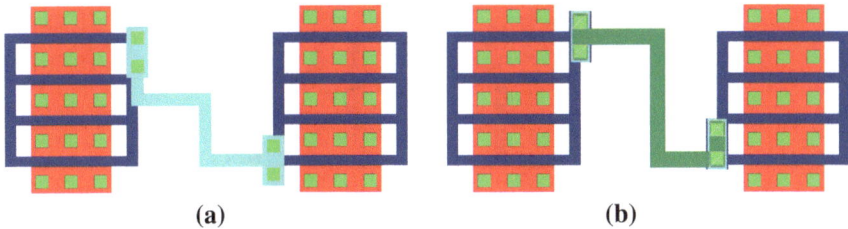

Fig. 5.9 Example of two possible parents. **a** Parent1 (P1), **b** Parent 2 (P2)

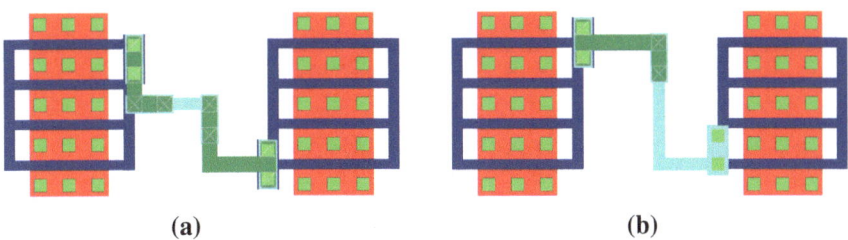

Fig. 5.10 Example of two possible offspring generated by crossover, after correction tasks. **a** Offpring 1 (O1′), **b** Offspring 2 (O2′)

connecting is randomly assigned. The exceptions are the source and sink contact orientations since those depend on the orientation of the device.

After the referred corrections, the offspring obtained is denoted as O1′ and O2′ in Table 5.2, and presented in Fig. 5.10.

5.3.2.4 Genetic Operator: Mutation

During mutation, a set of operators are applied to the wires according to the mutation ratio, which will introduce diversity in the population gene pool. They are classified in two categories: geometric operations and layer shifting operations. In the Fig. 5.11b–f different mutation operators were applied to the heuristic presented in Fig. 5.11a. These operations are applicable to any wire, and the geometric operations are the following:

- Randomly select a new source pin from the source terminal (Fig. 5.11b);
- Randomly select a new sink pin from the sink terminal;
- Randomize a new heuristic for the wire;
- Vertical slide on a horizontal segment;
- Horizontal slide on a vertical segment (Fig. 5.11c);
- Delete a vertical segment (Fig. 5.11d);
- Delete a horizontal segment;
- Randomize contact orientation (Fig. 5.11e).

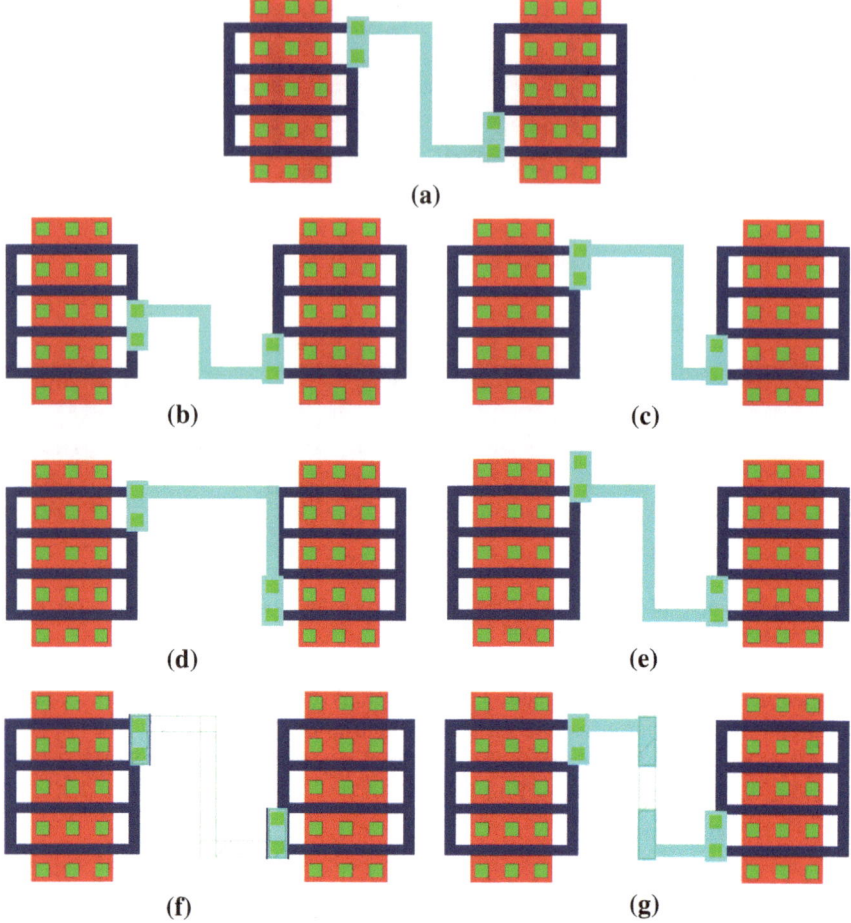

Fig. 5.11 Examples of mutation operators. **a** Initial heuristic of three segments, **b** New source pin, **c** Vertical slide, **d** Vertical segment deleted, **e** New contact orientation, **f** Move all line segments to a random conductor, **g** Move a line segment to a random conductor

The layer shifting operations are:

- Move all line segments to a random conductor (Fig. 5.11f);
- Move a line segment to a random conductor (Fig. 5.11g).

Since the router follows a greedy approach in the initialization, every wire is set to connect the two closest possible pins. The two change pins (source and sink) operators are essential to introduce some diversity in the population since the closest connection may not be physically achievable. Using all the available locations to connect the wires will result in better routings. As the randomize source and sink pin operators, the layer shifting operators are fundamental to circumvent the initial greedy approach where all wires from all nets were set to the same lowest cost conductor.

Deleting and sliding segments allow exploring endless combinations between wires geometries. These geometric operations allow using more effectively the lowest levels of conductors for a given technology, causing the wires adapt to each other without having the need of requiring higher level conductors to avoid short circuits and fulfilling design rules. The randomize heuristic operator exist to add fresh information to the gene pool, as the other operators are incremental, making less aggressive changes to the wire structure.

After each mutation, since numerous changes have been performed over each wire, the connectivity may have disrupted, therefore it is necessary to perform the full set of correction tasks used before. As in the initialization by heuristics, namely the *remove small segments* validation followed by the *rescale* step to ensure connectivity. And then as made after crossover, the *remove inverted segments* validation and finally the *verify contacts orientation* task.

As it is possible to observe in the Fig. 5.7, the initial heuristics always create wires confined in the rectangle defined between the two pins (edges). The genetic operators allow for out of bounding box exploration, this is, the wire geometry may depart as much as necessary away from the initial rectangle, as long as the connectivity is guaranteed.

5.3.3 Optimization Phases

The optimization process, presented in Fig. 5.1, is split in two optimization loops. Phase I and Detailed Routing represent the optimization stages of the router. The existence of two optimization stages, is due to the fact that when the exact physical representation of the wire is generated the number of layout elements being handled greatly increase, and the evaluation process becomes much slower because that its complexity is at best $O(n^2)$.

It was noticed that if the contacts, vias, and respective enclosing shapes were not generated, the evaluation was much faster. For this reason, in Phase I, which starts from the greedy initialization, only the main shape of each segment is generated and evaluated. This makes the processing faster and allows the exploration the general shape of the wires, obviously, the results might be infeasible, but they are obtained much faster. Both SCC, DRC as ERC validations are performed on each one of the phases.

Detailed Routing uses the output of phase I as initial population, and performs a more local and detailed optimization, taking into account the exact shape of the wire, considering every contact, via and conductor plates. When the solution is found, the generated routing is added to the placement and the target layout is saved in a GDSII file.

The routing for the template of Fig. 5.2c was automatically generated using the floorplan of Fig. 5.2b. Given that no net was marked as noisy or sensitive, the

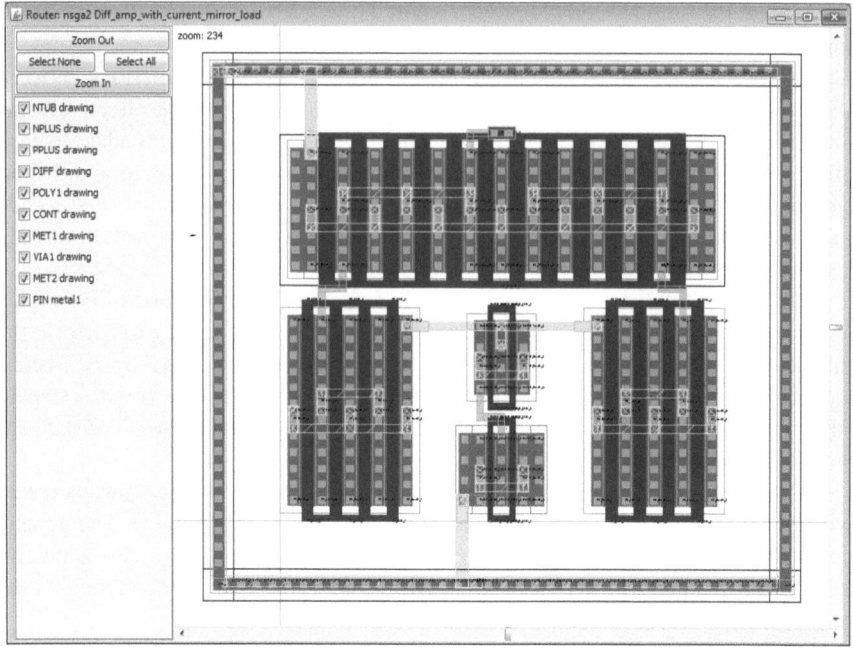

Fig. 5.12 Detailed routing obtained

optimization constraints considered were the number of short circuits and the number of design rules violations. The optimization objectives were the total wiring length computed together with the conductor cost associated and the number of contacts used. The layout was generated for a population of 128 elements, for 100 generations, along with a mutation rate of 3 % and a crossover rate of 90 %. For the addressed problem, each chromosome contains 9 genes (wires), and each wire is constituted by four linked segments. The layout was obtained in approximately 18 s, using the developed framework detailed in Chap. 6 and running on an Intel® Core™ 2 Quad CPU 2.4 GHz with 6 GB of RAM. The detailed routing obtained for the 0.13 μm UMC (United Microelectronics Corporation Group) design process is presented in Fig. 5.12.

As depicted in the figure bellow, each defined power net is automatically adjusted to the width or height of the guard ring, depending on the position defined by the designer in the guidelines. Since the guard ring is biased with the same potential of *vdd* power net, the stripe of conductor defining the power net is drawn with the same conductor.

5.4 Internal Evaluation Procedure

In order to evaluate efficiently the correctness of the circuit, an internal evaluation procedure was implemented that performs lightweight, but accurate design rule check, short circuits check, and some electrical rules check. The violations of the rules are used as constraints during optimization, and must be driven to zero.

To further increase the efficiency of the evaluation, it is done using one thread per available CPU, splitting among them the individuals being evaluated. Given the nature of evolutionary algorithms and the independence of the individual in the population the use of multi-threading in the evaluation procedure is greatly simplified. In addition, since nowadays most of the workstations have more than one CPU, a local thread pool can be used to accelerate the evaluation without many of complications associated with remote multi-threading or clustering parallelization techniques. For the current implementation, the use of three threads instead of a single thread, generated solutions in about 95 % smaller computational times for the evaluation of Phase I, and around 135 % smaller for the evaluation of Detailed Routing.

The use of an internal evaluation procedure rather than executing and then parsing the reports from an external tool, has many advantages when comparing the computational times. However, the commercial tool is more reliable than a custom made evaluator. This internal evaluation is constantly subjected to an exhaustive debug to ensure that the validations are being made correctly. The detection of false errors may difficult the convergence of the optimization kernels. The reliability is not compromised since the results will be either way validated in Calibre® DRC tool.

5.4.1 Short Circuit Checker

The way the wires are operated during the optimization ensures the desired connectivity is present in the output, however is does not ensure that unwanted connections between nets are not created. The short circuit checker is used to count the number of unwanted connections between nets, and this must be reduced to zero during generation. The pseudo code for the SSC is presented in Fig. 5.13.

For each wire in the layout, its' shapes are checked against all the shapes from the devices (except the ones connected to the wire source/sink) and from other nets. Only shapes in the metals and poly layers are checked, because when contacts or vias are instantiated, the enclosing metal shapes (above and below) are always added.

```
short_circuits = 0;
for (each Wire W in Layout) {
List Wire_shapes = Wire.get_Physical_Representation;
    //Check for Short Circuits with Placement
    for (each different Layer L in Wire_shapes)
            List Layer_shapes_wire = Wire_shapes.get(Shapes from Layer L);
            List Layer_shapes_placement = Placement.get(Shapes from Layer L);

            Layer_shapes_placement.remove(Wire W Source Shapes);
            Layer_shapes_placement.remove(Wire W Sink Shapes);

            for (each shape sh1 in Layer_shapes_wire)
                    for (each shape sh2 in Layer_shapes_placement)
                            if (sh1 intersects sh2)
                                    short_circuits ++;
                                    add markers in layout viewer;

    //Check for Short Circuits with other Nets
    for (each other Net N not containing Wire W)
            for (each different Layer L in Wire_shapes)
                    List Layer_shapes_wire = Wire_shapes.get(Shapes from Layer L);
                    List Layer_shapes_net = Net N.get(Shapes from Layer L);

                    for (each shape sh1 in Layer_shapes_wire)
                            for (each shape sh2 in Layer_shapes_net)
                                    if (sh1 intersects sh2)
                                            short_circuits ++;
                                            add markers in layout viewer;
}
```

Fig. 5.13 Short circuit check algorithm

5.4.2 Design Rule Checker

To ensure that the generated routing complies with the process design rules, a mixture of correct by construction and optimization is used. Some rules, like the enclosing of vias by metals or minimum sizes are not validated, since they are generated in a correct-by-construction method. All shapes that are created within the application must be correct, therefore, if during the Calibre® DRC verification any of these rules is violated, it means that there is a bug in the generation code or design kit, not in the verification procedure.

However, in the way the problem is modeled, some rules may be violated during the generation, and must be verified and their occurrence must be driven to zero. The rules that are verified using the internal design rule checker are:

- Minimum spacing between conductors;
- Minimum spacing between contacts or vias;
- Maximum widths of conductors, contacts and/or vias.

In some technology design kits, the vias have a fixed size that must be verified because even though different shapes are created with the correct size, overlapping two vias may create an illegal shape (only if they totally overlap the layout is correct).

Each segment or shape as an associated bounding box. The minimum distances validation procedure is essentially an interception check between the expanded bounding box of the actual shape and every other shape in the same conductor. The initial bounding box was expanded in each direction by a size equal to the minimum distance allowed by target technology. However, a shape may have valid connections to any number of other shapes and still has to verify a fixed distance with the remaining shapes. Obtaining this special expanded bounding box is where the main computational effort of design rule check lies.

Because the layout of the placed devices is correct, those shapes do not need to be verified, however for each wire all its' shapes should be verified against all the placement shapes and the shapes from the other wires. The verification approach used was to iteratively add the shapes of each wire to the layout and compute the number of violations when comparing the wire shapes against each other, and against the shapes already present in the layout.

5.4.3 Electrical Rule Checker

In the current implementation, the electrical rule checker is used essentially to identify the constraints violated by the noisy and sensitive signals, as detailed next. As the tool is being adapted for smaller design processes, the treatment of other important electrical events must be taken into account.

In an AMS ICs, analog signals are far more-noise sensitive than digital counterpart. The most sensitive analog signals are those that carry very low-level signals at high impedance levels, for example [6]: inputs to high-gain amplifiers, precision comparators and analog-to-digital converters; outputs of precision voltage references; analog ground lines to high-precision circuitry; and low-level signals at high impedance. Usually, noisy and sensitive signals are identified during the specification translation task to be properly treated during the routing task, in which exist a set of common techniques used by designers. The identification of these lines is a difficult task in the absence of electrical measures, and most layout designers do not possess the knowledge and experience required to correctly identify all of them in a complicated analog circuit. As stated for LAYGEN II, designer uses the template to identify all the noisy and sensitive nets, and the router deals automatically with them.

Typically noisy circuitry is placed as far away as possible from sensitive circuitry. It is known that noisy signals should not run on top of sensitive signals, but if a crossing is required, the area of intersection should be reduced as possible and an electrostatic shield technique used. A practical method of constructing an electrostatic shield is to run one signal in one conductor and the other signal in a conductor two levels higher. A plate of conductor connecting to a low-impedance node, e.g., ground line, is interposed in the conductor between the two signals and acts as a shield. Also, in order to run noisy signals adjacently to sensitive, a shield line must be used between the two. The shield line used is generally a low-noise, low-impedance signal such as ground line, supply line or even the output of digital logic gates [6].

Table 5.3 Summary of constraints and objectives

Constraint	Target	Description
SCC	= 0	Short circuits
DRC	= 0	Technology design rules violations
ERC	= 0	Crossing between noisy and sensitive nets, or running on top of devices
Objective	Target	Description
Wiring lengh	minimize	Total wiring length computed with the associated conductors cost
Contacts	minimize	Number of contacts or vias used
Distance	maximize	Distance between noisy and sensitive nets

While the techniques listed can somehow be used with more or less success in the manual design, from the standpoint of automatic generation the use of extra lines connected to ground or power, either to perform electrostatic shields as shield lines, would mean an abrupt increase of the solution's complexity. As LAYGEN II's router uses the wiring length as objective, is ensured that the sensitive signals are always as short as possible, reducing the opportunities for noise coupling. To assure the best possible treatment for the nets denoted as noisy or sensitive without unduly increase the solution complexity, any crossing between any wire of a net denoted as noisy and another denoted as sensitive is considered an ERC error.

It is know that sensitive signals should not pass through other circuit blocks, the same can be applied to noisy signals if the sensitive devices of a given circuit block where not correctly identified. In order to reduce the possible capacitive coupled noise in the optimization, each noisy and sensitive net running on top of the active area of the devices is marked as an ERC violation. Simultaneously with the verification of the possible interceptions, it is computed the minimum distance between any segment of a noisy line and a sensitive one. This computed distance may be used as objective during the evolutionary optimization in order to maximize it, and place the two types of nets as far away as possible.

The internal evaluator also marks individually each wire as he contains any constraint violation or not. In the next generation and subsequent mutation, wires that fulfill all the constraints have a smaller mutation ratio than the wires violating constraints. This way the wires without errors suffer fewer changes and are preferably kept in the solutions, the ones containing errors need to evolve to adapt to the remaining and become valid.

5.5 Conclusions

In this chapter, the optimization-based router was described and the simple amplifier presented in the previous chapter was used to demonstrate the techniques applied. A summary of the constraints and objectives used by the MOEA are

5.5 Conclusions

depicted in Table 5.3, other constraints or objectives can be easily added as suited for the designer. The optimization algorithm and hence the chromosome structure and genetic operators were properly detailed. Along with the internal evaluation procedure that allows LAYGEN II to avoid the need of external evaluation tools.

Unlike the placer where a strict template-based approach is followed, in the router, LAYGEN II replaces the obligation to describe the exact routing, by a set of constraints that guide the automatic generation to solution according to the designer requirements. The routing constraints are independent of the floorplan and can be valid even for different placement topologies. Plus, the connectivity is immutable no matter the changes done in the floorplan, since they only depend of the circuit.

This extremely versatile approach, allows for the designer to provide the connectivity, and a set of symmetry and sensitivity constraints, and the tool automatically starts the optimization procedure that will lead to a solution that strictly complies with the target technology design rules. The solution is generated and validated even if only connectivity is provided.

References

1. N. Lourenço, M. Vianello, J. Guilherme, N. Horta, LAYGEN—Automatic layout generation of analog ICs from hierarchical template descriptions, in *Proceedings Conference on Ph.D. Research in Microelectronics and Electronics (PRIME)*, pp. 213–216 (2006)
2. F. Maloberti, *Analog Design for CMOS VLSI Systems* (Kluwer Academic Publishers, Boston, 2001)
3. A.E. Eiben, J.E. Smith, *Introduction to Evolutionary Computing* (Springer, Berlin, 2003)
4. K. Deb, A. Pratap, S. Agarwal, T. Meyarivan, A fast and elitist multiobjective genetic algorithm: NSGA-II. IEEE Trans. Evol. Comput. **6**(2), 182–197 (2002)
5. Mentor Graphics, http://www.mentor.com
6. A. Hastings, *The Art of Analog Layout*, 2nd edn. (Prentice Hall, New Jersey, 2005)

Chapter 6
Results

Abstract This chapter illustrates the application of the proposed design flow to practical examples. In order to use the implemented platform, the first design task is the development of the technology design kits. The 0.13 μm UMC (United Microelectronics Corporation Group) complementary metal–oxide–semiconductor (CMOS) design kit presented in the previous chapters will be used for the design of most of the layouts presented in this chapter. Since its design accompanied the development of the tool, it is difficult to report the development time, but the development of a similar design kit, e.g., 180 or 130 nm from another vender, can be achieved in less than 2 weeks. The framework of the proposed methodology for the automatic generation of analog ICs layout has been coded in JAVA and is running, for the two presented examples, on an Intel® Core™ 2 Quad CPU 2.4 GHz with 6 GB of RAM, three threads are being used to run the evaluation procedure of each population, at each generation of the router. The code automatically generates the GDSII file required by Mentor Graphics' Calibre® [1] tool.

Keywords Analog IC design · Automatic layout generation · Evolutionary computation · Layout retargeting · Physical design · Technology migration

6.1 Case Study I: Fully-Dynamic Comparator

The first circuit used as test case is a fully-dynamic comparator which is part of a ΔΣ Modulator [2], courtesy of Prof. João Goes from CTS-UNINOVA. The comparator schematic is presented in Fig. 6.1 and was designed for a 130 nm CMOS technology. The handmade layout of Fig. 6.2 was designed for the devices sizes provided in Table 6.1.

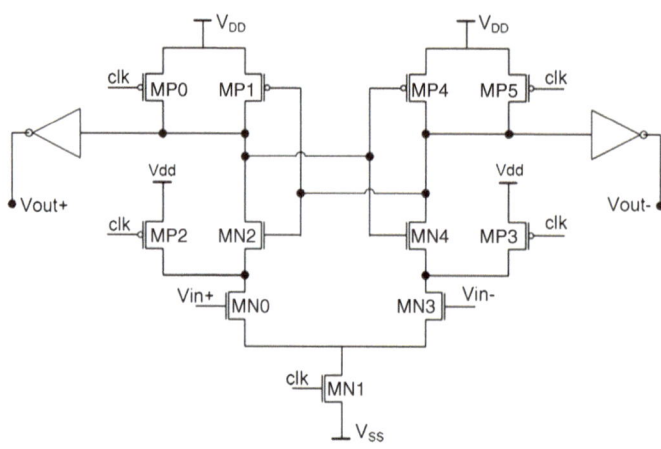

Fig. 6.1 Electrical schematic of the fully-dynamic comparator

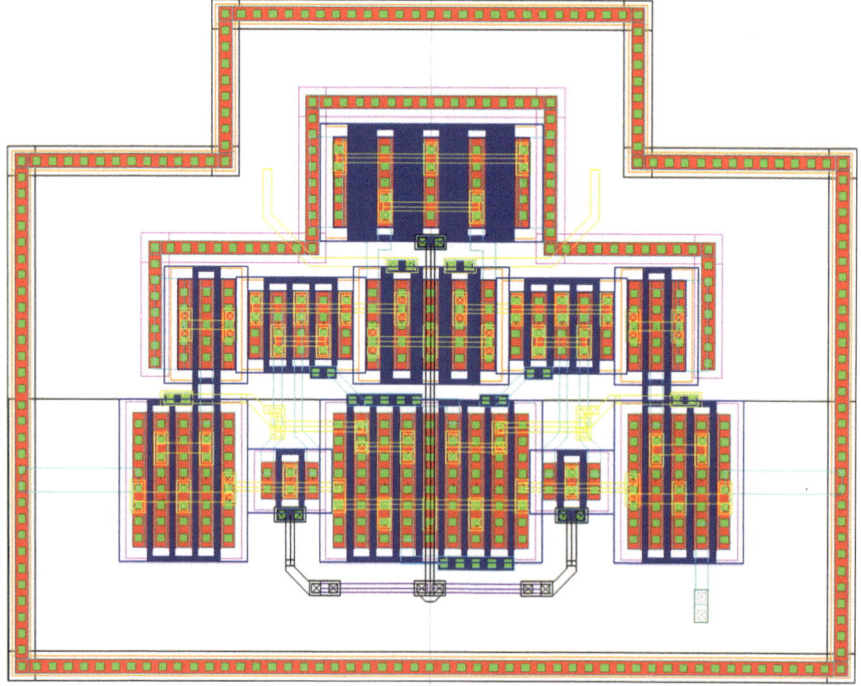

Fig. 6.2 Comparator handmade layout (0.13 μm design process)

6.1 Case Study I: Fully-Dynamic Comparator

Table 6.1 Comparator sizing (0.13 μm design process)

Devices	Width (μm)	Length (nm)	Gates
Inverter MN0	4	120	2
Inverter MP0	12	120	4
MN0, MN3	4	260	2
MN1	8	640	4
MN2, MN4	6	120	4
MP0, MP2, MP3, MP5	0.820	120	1
MP1, MP4	12	120	4

6.1.1 Template

The proposed design flow starts with the template's definition. To incorporate design strategies used by the designer when performing handmade layout, the template guidelines were derived from the handmade layout of Fig. 6.2. Besides being used to extract the expert knowledge, the handmade layout also provided an evaluation measure for the quality of the target layout. LAYGEN II inputs the design-kit and the template, and is used to automatically generate the target layout.

The first step in defining the template is to identify possible inner-templates. In this circuit, the hierarchy partitioning used is the one depicted in Fig. 6.3. The partition 1 includes all the PMOS transistors with sources connecting to *vdd* potential, while partition 2 includes the remaining NMOS transistors with the exception of transistor MN1, which was moved to top cell for symmetry purposes. Guard ring is only requested for the top cell.

The routing connectivity was equally provided, and all nets suited to be symmetric were marked as such. The *clock* was labeled as noisy and, without using any measures that proved it, the *gnd* connection and the connections from transistor MN1 to transistors MN0 and MN3 were marked as sensitive. The previous attributions were used only to identify the behavior of the optimization kernel in the presence of special nets. Also, two power nets for *vdd* potential, placed bellow the circuit and *gnd* potential, placed above were defined. All necessary routing of the sub-templates and intra-templates is performed in the top cell, in total the connectivity for 36 wires was provided, distributed by nine nets.

6.1.2 Layout Generation

First all placement tasks are performed, from the inner templates to the top cell, and only then the routing tasks are executed.

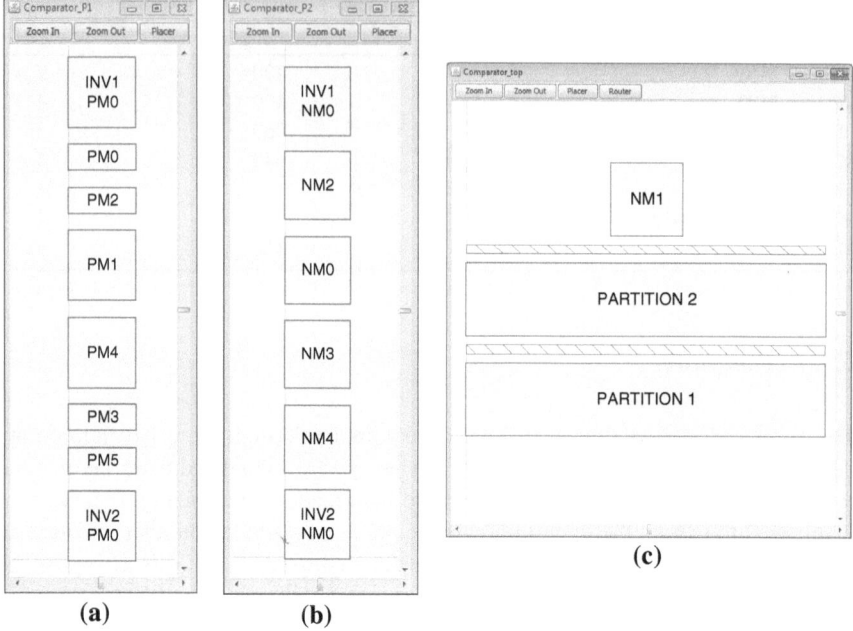

Fig. 6.3 Comparator template hierarchy. **a** Partition 1. **b** Partition 2. **c** Top partition

6.1.2.1 Placer

Using the template partitioning of Fig. 6.3 and the parametric module generator of the 130 nm technology design kit, the placer starts by generating the inner templates first. The layout generated for partition 1 is shown in Fig. 6.4a, while partition 2 in Fig. 6.4b and, finally, the top cell in Fig. 6.4b, which is placed only after all the sub partitions are available. All the pre and post processing task are performed.

In partition 1, transistor pairs PM0 and PM2, PM1 and PM4, and also PM3 and PM5 were automatically merged; and for the partition 2 the pair NM0 and NM3. The placement is totally symmetric, even the biasing considerations included in the transistors. Although partition 1 and 2 were firstly generated vertically, since the layout representation structure only allow for vertical symmetry axis, the partitions were then rotated to fit correctly in the top cell. A guard ring with N-Well was automatically adjusted to the obtained floorplan.

6.1.2.2 Router

The floorplan of Fig. 6.4c is used as starting point for routing optimization. The optimization kernels have a population of 256 elements, both the Phase I and the Detailed Routing were optimized for 200 generations. The convergence of the

6.1 Case Study I: Fully-Dynamic Comparator

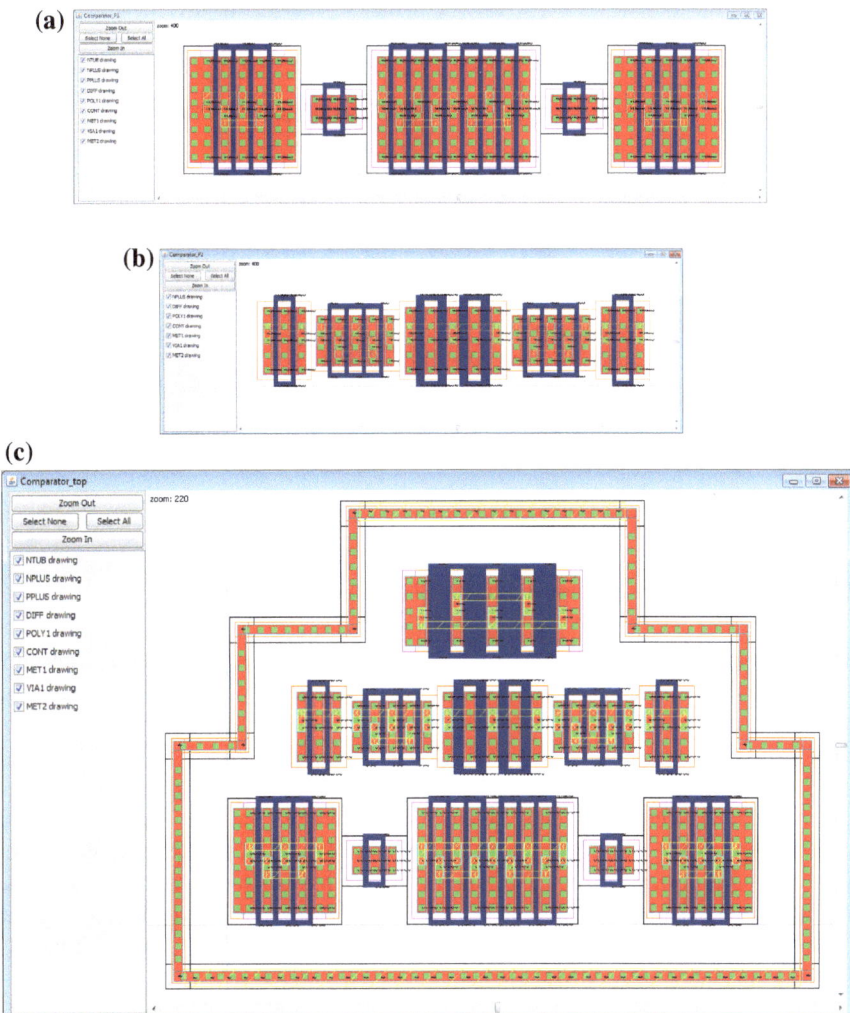

Fig. 6.4 Floorplans obtained (0.13 μm design process). **a** Partition 1. **b** Partition. **c** Top cell

algorithm strongly depends on the greedy initializations performed for each element. So, populations with higher number of elements present more diversity and consequently fewer generations required to achieve feasibility. For complex routing problems, the number of elements should be kept above the 128 elements, allied to a considerable number of generations. As characteristic of the current algorithm, the objectives are only taken into account when all the constraints are fulfilled, so there has to be ensured that the evolutionary kernel optimize the solutions for a substantial number of generations after the feasibility is attained.

The results were generated for a mutation rate of 3 % and a crossover rate of 90 %. The mutation rates for the current optimization problem superior to 5 %

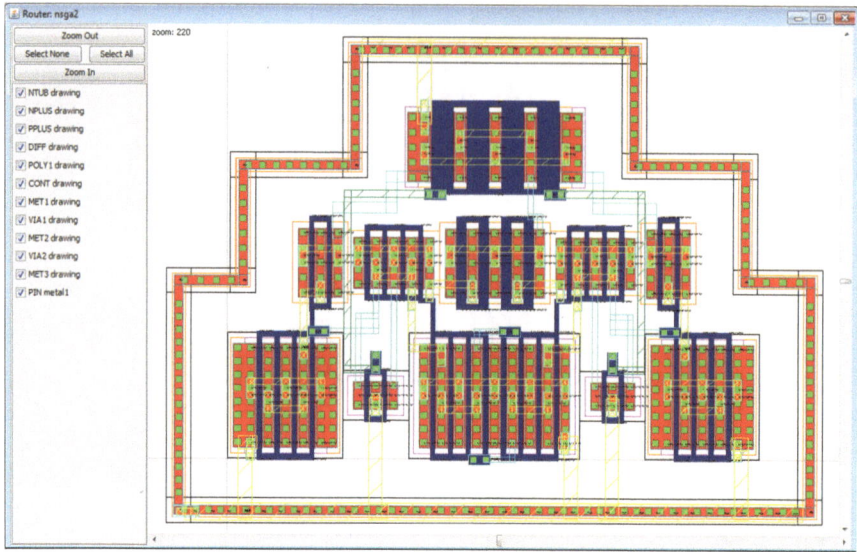

Fig. 6.5 Automatically generated layout (0.13 μm design process)

proved to disperse the elements of the population, which hinders the algorithm convergence, so this value need to be kept significantly low. For its part, the crossover rates can be raised up to 100 % without deteriorating the performance of the algorithm.

Four constraints were used in the optimization, namely short circuits, design rule violations, illegal crossing between noisy and signal, and finally noisy or signal nets running on top of the active area of the devices. The three objectives used were minimizing the wiring length considering the conductor cost associated to each segment, minimizing the number of contacts used and maximizing the distance between noisy and signal nets. The obtained solution and consequently final layout is presented in Fig. 6.5.

Table 6.2 summarizes the execution times for all the template hierarchy of the comparator, it is easy to see that the placer execution times are almost instantaneous when compared to the router, which dominated more than 99 % of the computation time. The automatic generation times (approximately 281 s) are obviously negligible when compared to the manual design, for a fair comparison is necessary to estimate the complete design of a circuit, which encompasses the template setup and guidelines adjustment times. The complete design of this fully-dynamic comparator is difficult to measure since his development accompanied the development of the tool. The new implementations have been iteratively integrated and tested in the current circuit, leaving the considerations about complete design times to the next test case presented in this chapter.

A direct loss when compared to the manual design is the lack of 45° wires. Although they might not be essential in the current circuit for a 130 nm process,

6.1 Case Study I: Fully-Dynamic Comparator

Table 6.2 Execution times summary

Template	Placement time (ms)	Routing time		Total
		Phase I	Detailed	
Partition 1	83	Not performed		83 ms
Partition 2	39	Not performed		39 ms
Top partition	50	97.571 s	183.286 s	280.907 s

for radio frequency circuits and smaller design processes the 45° wires are a particular request of the designers, so they are therefore addressed as a future implementation in the proper section of Chap. 7. Nevertheless it does not invalidate the conclusions about the validity of the proposed design approach.

It is important to notice the symmetry nets in the obtained layout and also many of the initial greedy considerations were kept. Only the three first levels of metals from the eight available in the current technology were used in the generated solution.

6.1.3 Validation

The GDSII file was generated for the top cell and the results were successfully validated by Calibre® [1] design rule check (DRC) tool. Once the DRC verification is concluded, the next step is to verify if the layout matches the electrical schematic, using Calibre® layout versus schematic (LVS). With the DRC and LVS verified, it is necessary to create an extracted view of the layout including the parasitic elements, e.g., resistances/capacitances of layout traces and coupling capacitances, to perform post-layout simulations. Transient simulations of the electrical schematic, extracted netlist from the handmade layout, and extracted netlist from the automatically generated layout were performed and the outputs are presented in Fig. 6.6.

The presented result is the output of the comparator for a state change in a rising edge of the clock. It is relevant to notice that both handmade and automatically generated layouts present similar response times, however, it is not intended to take any conclusions about performances. The objective was to prove the use of LAYGEN II to automatically generate layouts, that can be validated in an industrial grade DRC and LVS tool, and successfully perform post-layout simulations. The achieved extracted comparator would require a complete characterization, in order to confirm the performance results of all specifications of the original design. This characterization is out of the scoop of this book.

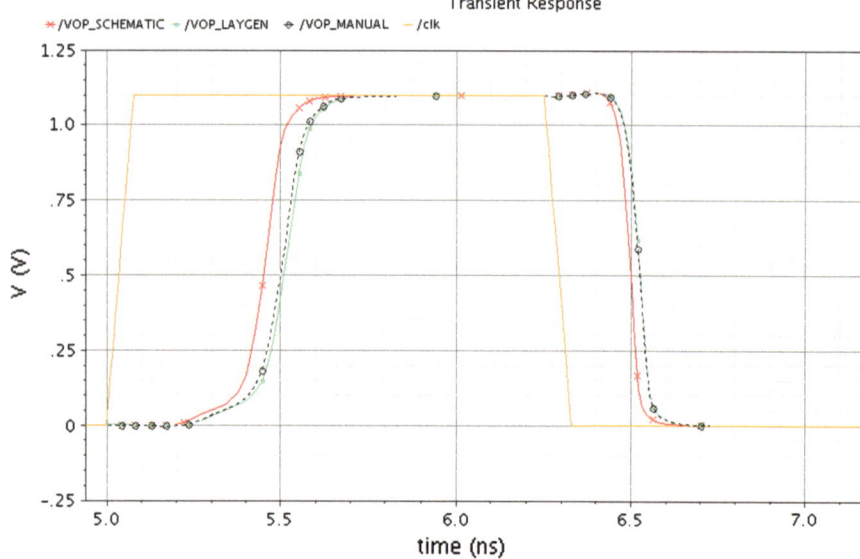

Fig. 6.6 Comparator simulation: schematic, handmade, and LAYGEN II layout

Fig. 6.7 Electrical schematic of the single-ended folded cascode amplifier

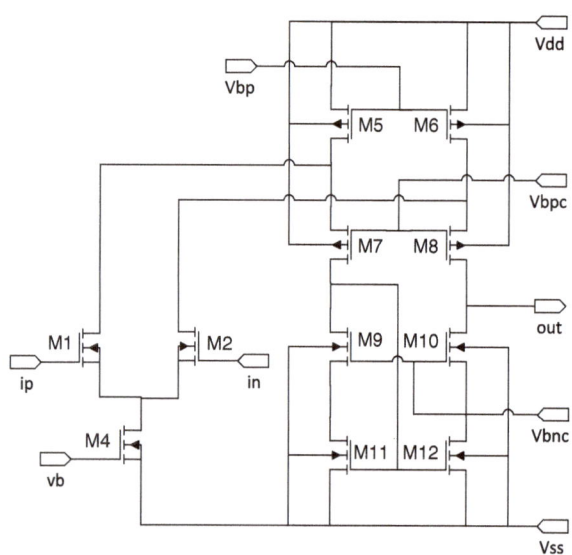

6.2 Case Study II: Single-Ended Folded Cascode Amplifier

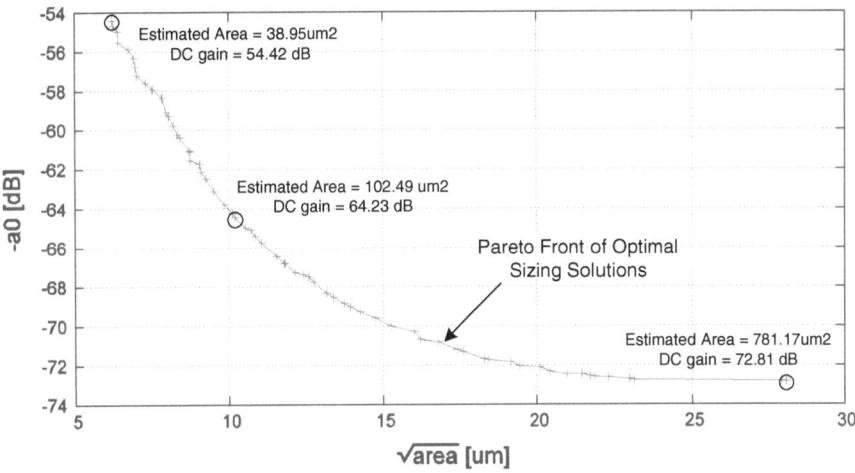

Fig. 6.8 POF obtained during sizing task, gain [dB] versus estimated area [μm^2]

6.2 Case Study II: Single-Ended Folded Cascode Amplifier

The second case study of this chapter is a single ended folded cascode operational amplifier (OPAMP) tested with FRIDGE synthesis tool [3], the circuit schematic is shown in Fig. 6.7.

A previous sizing task was performed by GENOM-POF [4–7] whose optimization objectives were minimizing the area and maximizing the gain. The gain versus area Pareto optimal front (POF) obtained is presented in Fig. 6.8 and three sizing solutions were selected. The sizing solution which minimizes presents the minor area was selected as the first point from the obtained POF. A second solution was randomly selected from the remaining solution space, with the only restriction that the area was substantially greater than the area of the first solution. The third and last solution is the solution that presents the large gain. Those sizing solutions are presented in Table 6.3.

6.2.1 Template Hierarchy

The first set of device sizes of Table 6.3 was used to define the guidelines of the floorplan. The partition 1 encompasses all the PMOS transistors of the circuit; partition 2 includes two PMOS of the cascode, while partition 3 corresponds to the differential pair. Partition 2 is instantiated twice since the cascode requires two plus two symmetric NMOS transistors. The hierarchy partitioning used is the one depicted in Fig. 6.9.

Table 6.3 Measures and devices sizes attained during sizing task for the amplifier, using GENOM-POF for the 0.13 μm UMC design process

Measures	Devices	Width (μm)	Length (nm)
Estimated area = 38.95 μm^2	M1, M2	14.67	480
DC gain = 54.42 dB	M4	3.84	530
	M5, M6	13.4	140
	M7, M8	17.77	370
	M9, M10	5.72	310
	M11, M12	2.53	470
Estimated area = 102.49 μm^2	M1, M2	13.2	490
DC gain = 64.23 dB	M4	11.33	400
	M5, M6	27.29	290
	M7, M8	36.86	530
	M9, M10	17.61	540
	M11, M12	7.61	730
Estimated area = 781.17 μm^2	M1, M2	41.22	760
DC gain = 72.81 dB	M4	69.09	670
	M5, M6	153.37	570
	M7, M8	249.84	780
	M9, M10	55.29	790
	M11, M12	12.21	800

The routing connectivity was equally provided, and all nets suited to be symmetric were marked as such, and two power nets for *vdd* and *gnd*, placed above and below the circuit. All necessary routing of the sub-templates is performed in the top cell, in total the connectivity for 25 wires was provided, divided into 12 different nets.

6.2.2 Layout Generation

As before, first all bottom-up placement tasks are performed and then routing task is executed. Using the template partitioning of Fig. 6.9 and the parametric module generator of the 130 nm technology design kit, the placer starts by generating the inner templates first. The layout generated for partition 1 is shown in Fig. 6.10a, while partition 2 in Fig. 6.10b and partition 3 in Fig. 6.10c. The layout of the top partition is presented in Fig. 6.10d, which is placed only after all the sub partitions are available. All the pre and post processing task are performed.

In partition 1 transistor pairs M5 and M6 were automatically merged, and for the partition 3 the pair M1 and M2. The placement obtained is totally symmetric and a guard ring with N-Well was automatically adjusted to the obtained floorplan.

The floorplan of Fig. 6.10d is used as starting point for routing optimization. The optimization kernels have a population of 128 elements, both the Phase I and the Detailed Routing were optimized for 200 generations. The results were

6.2 Case Study II: Single-Ended Folded Cascode Amplifier

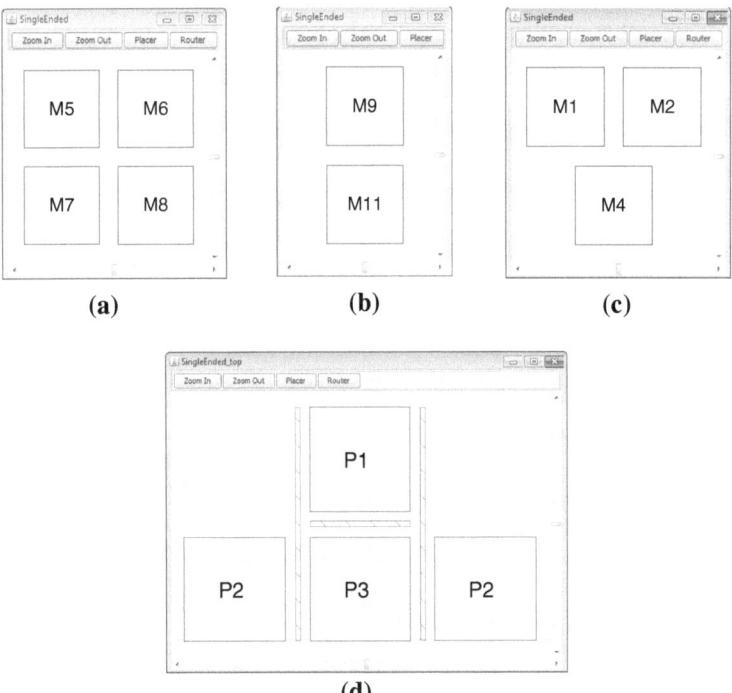

Fig. 6.9 Amplifier template hierarchy. **a** Partition 1. **b** Partition 2. **c** Partition 3. **d** Top partition

generated for a mutation rate of 3 % and a crossover rate of 90 %. The obtained solution, i.e. the final layout, is presented in Fig. 6.11. Table 6.4 summarizes the execution times for all the templates in the hierarchy of the circuit. Two constraints were used in the optimization, namely short circuits and design rule violations, the two objectives used were minimizing the wiring length considering the conductor cost associated to each segment and the number of contacts used.

The total template description was done in approximately 2 h, including the iterative process of refining the guidelines and the automatic generation was performed in less than 112 s. The presented results were successfully validated in Calibre® DRC tool.

6.2.3 Retargeting for Different Sizes

The low-level design flow proposed in this work is intended to increase the design reusability, without loss of the designer expertise, but also without overwhelming the designer. In this section the same template was used, but with different sizes for the devices. The new sizes are the second solution presented in Table 6.3. This example took approximately the same computational time than the previous

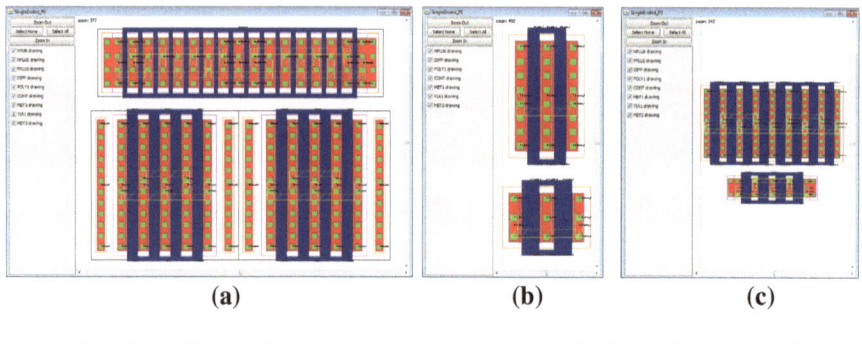

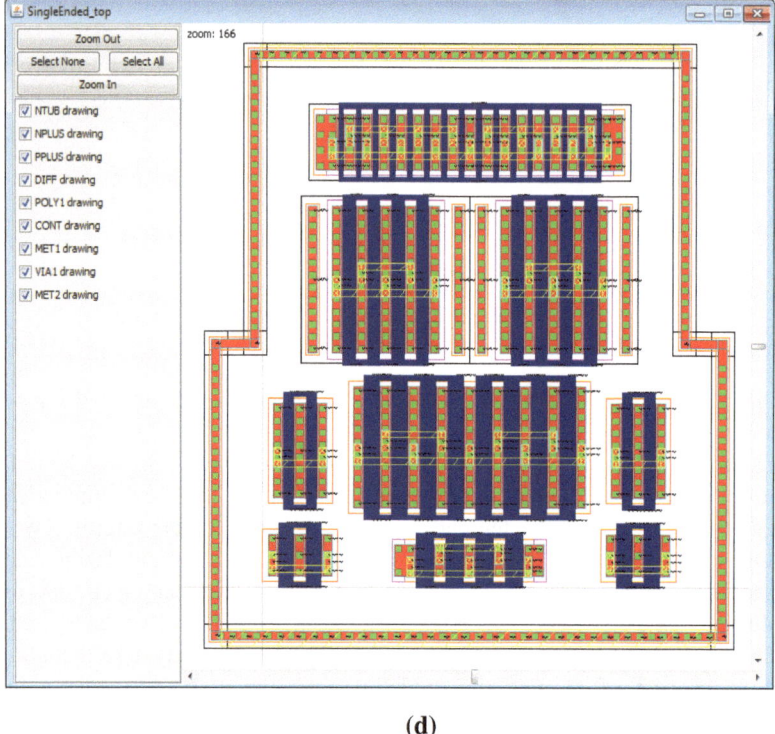

Fig. 6.10 Floorplan obtained for the top partition (0.13 μm design process). **a** Partition 1. **b** Partition 2. **c** Partition 3. **d** Top partition

example to be generated and the result is presented in Fig. 6.12b. The retargeting operation took less than 15 min to complete, including all the tasks: tweaking of the template and synthesis. The only change in the template was made to reflect the new sizes of the modules; no changes were performed in the routing template (connectivity and constraints).

In Fig. 6.13b is presented the retargeting for the third sizing solution of Table 6.3, which represents a huge change in the devices sizes comparatively to

6.2 Case Study II: Single-Ended Folded Cascode Amplifier

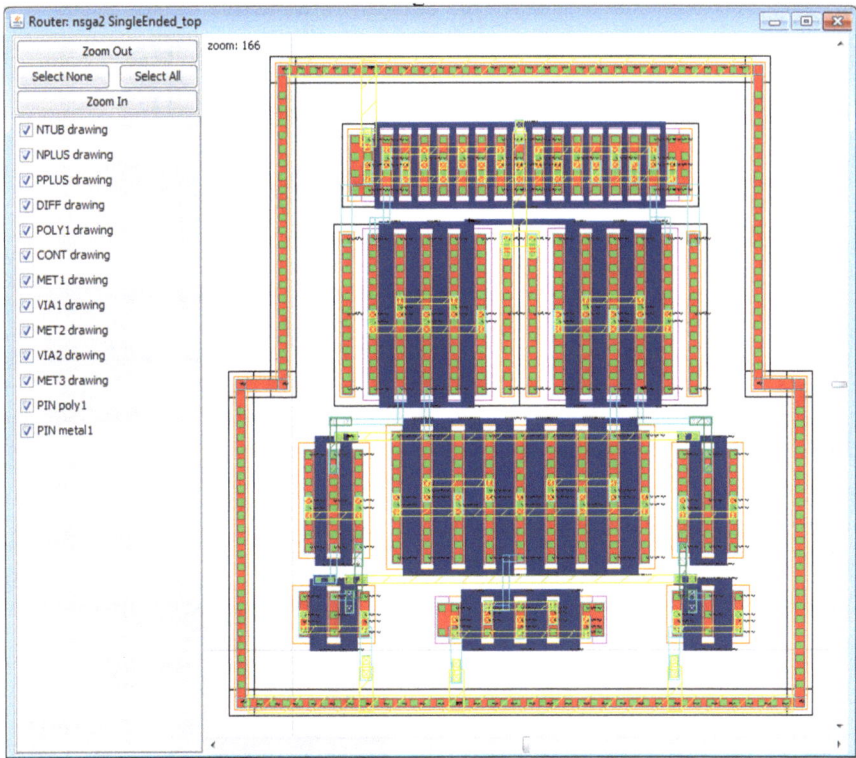

Fig. 6.11 Automatically generated layout for the first sizing solution (0.13 μm design process)

Table 6.4 Execution times summary

Template	Placement time (ms)	Routing time		Total
		Phase I	Detailed	
Partition 1	64	Not performed		64 ms
Partition 2	16	Not performed		16 ms
Partition 3	15	Not performed		15 ms
Top partition	41	40.438 s	71.060 s	111.539 s

the other solutions. The layouts automatically generated for the two previous sizing solutions are placed on the same scale in Fig. 6.13a. Again, no changes were performed in the template except the devices' sizes.

Considering that no change was performed in the high level floorplan, connectivity or constraints, for any of the retargeting operations above, the results are promising. However, the template was optimized for the first sizing solution and different topological relations between cells could be more suited for the remainder sizing solutions. The same template may not yield the best topological

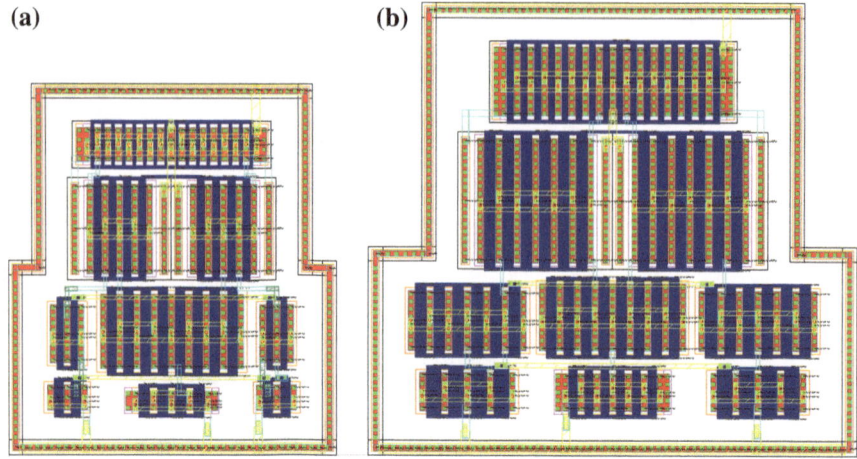

Fig. 6.12 Amplifier retargeting (0.13 μm design process). **a** Automatically generated layout for the first sizing solution. **b** Automatically generated layout for the second sizing solution

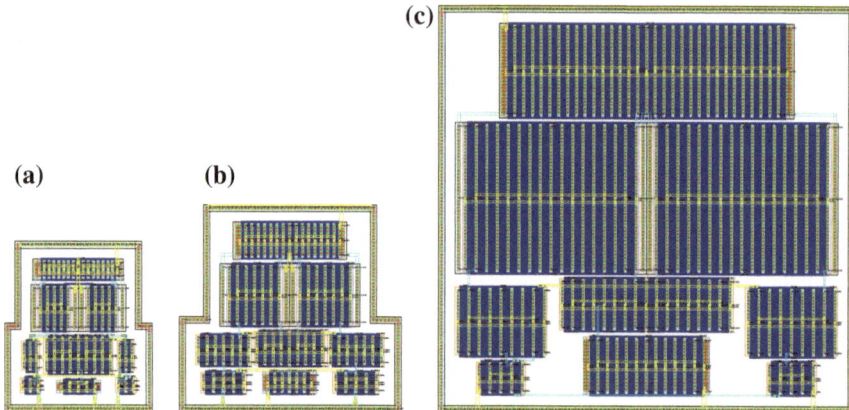

Fig. 6.13 Amplifier retargeting (0.13 μm design process). **a** Layouts of the first and second sizing solutions. **b** Automatically generated layout for the third sizing solution

Table 6.5 Measures and devices sizes attained during sizing task for the amplifier, using GENOM-POF for the 0.35 μm AMS design process

Measures	Devices	Width (μm)	Length (nm)
Estimated area = 46.79 μm^2	M1, M2	20	520
Power = 0.347 mW	M4	12	350
DC gain = 49.175 dB	M5, M6	8	420
	M7, M8	11	410
	M9, M10	2	480
	M11, M12	2	870

6.2 Case Study II: Single-Ended Folded Cascode Amplifier

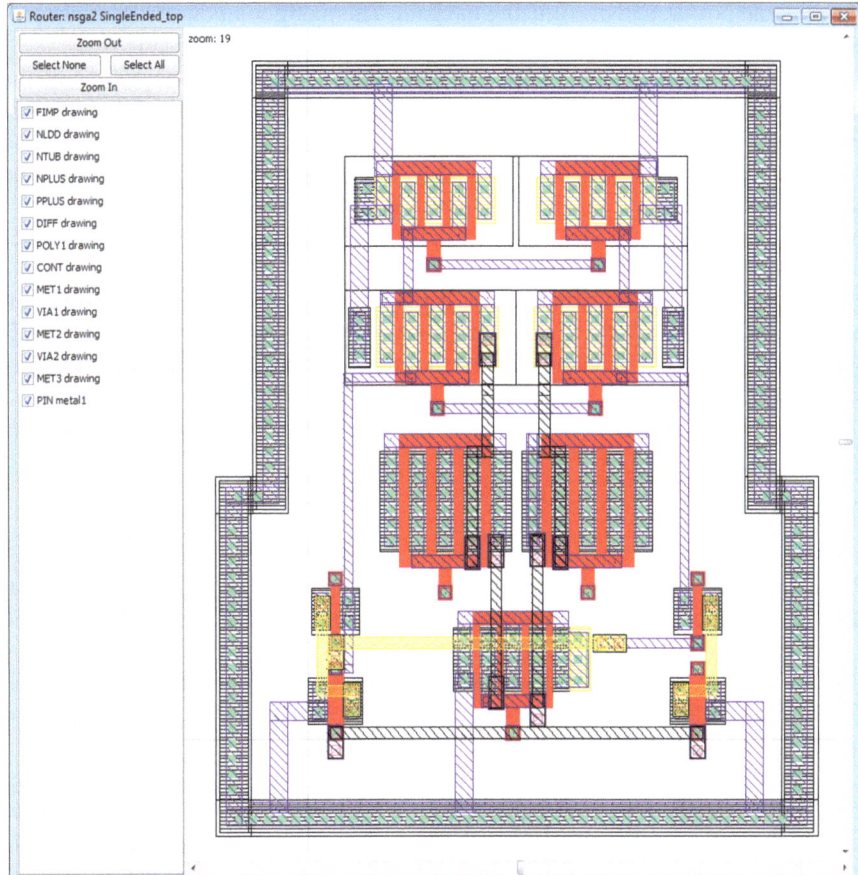

Fig. 6.14 Amplifier retargeting for different technology (0.35 μm design process)

relations between cells as devices' sizes are changed, thought as demonstrated, the connectivity is always valid.

In the following section, this template is used also for retargeting, but to a different technology.

6.2.4 Retargeting for Different Technology

The ability to support multiple technologies is mandatory if one attempts to achieve the maximum flexibility on retargeting operations. To demonstrate the technology independence of the proposed design flow, the template is going to be retargeted for the 0.35 μm AMS (Austria Micro Systems) design process. A previous sizing task was performed by GENOM-POF whose optimization objectives were minimizing the area and power, and maximizing the gain. The sizing solution which minimizes the area was selected from the obtained POF, and is presented in Table 6.5.

The only changes in the templates were made to reflect the new sizes of the modules, and the template's file header changed to the designation of the different technology. No changes were performed in routing template and this retargeting task was performed in less than 10 min. This process obviously assumes that the desired technology design kit is available. If available, the migration process is actually pretty simple. The obtained result from this retargeting operation is presented in Fig. 6.14.

Although the technology design kit is outdated when compared to the 0.13 μm technology design kit, lacking merged transistors and symmetry in all devices, it does not invalidate the obtained results to prove the concept. The presented results were successfully validated in Calibre® DRC tool.

6.3 Conclusions

In this chapter, two test cases were addressed to show the capabilities of the tool. The first example, a fully-dynamic comparator, was used to compare the LAYGEN II results with a handmade layout and perform a set of validations (DRC, LVS and post-layout simulation). The second example, a single-ended folded cascade amplifier, was used to explore the retargetability characteristics of the proposed methodology.

In the first test case the template was easier to design, because it was only required to translate the information from the handmade layout to the template. Designing the template from scratch and refining the high level floorplan is obviously a more time consuming task. When compared to manual design, the use of LAYGEN II implies some additional initial work to setup the template. However as shown in the second example, after the template is available the proposed design flow highly increases the reusability of the design.

The design tasks that are more time consuming are the development of the technology design kits. A retargeting for the 0.35 μm AMS design process was performed to show the versatility of technology migrations. Once the template and the target technology design kits are available, technology migration processes are performed for the same or different specifications within minutes of computation time.

References

1. Mentor Graphics, http://www.mentor.com
2. B. Nowacki, N. Paulino, J. Goes, A 1.2 V 300 μW second-order switched-capacitor $\Delta\Sigma$ modulator using ultra incomplete settling with 73 dB SNDR and 300 kHz BW in 130 nm

CMOS, in *Proceedings of the European Solid-State Device Research Conference (ESSCIRC)*, pp. 271–274 (Out 2011)
3. F. Medeiro, F. V. Fernandez, R. Dominguez-Castro, A. Rodriguez-Vazquez, A statistical optimization-based approach for automated sizing of analog cells, in *Procedings of ACM/IEEE International Conference on Computer-Aided Design (ICCAD)*, pp. 594–597 (November 1994)
4. M. Barros, J. Guilherme, N. Horta, GA-SVM feasibility model and optimization kernel applied to analog IC design automation, in *Proceedings of ACM Great Lakes symposium on VLSI (GLVLSI)*, pp. 469–472 (March 2007)
5. M. Barros, J. Guilherme, N. Horta, *Analog circuits and systems optimization based on evolutionary computation techniques*, Studies in computational intelligence, vol. 294 (Springer, New York 2010)
6. M. Barros, J. Guilherme, N. Horta, Analog circuits optimization based on evolutionary computation techniques. Integration, VLSI J. **43**(1), 136–155 (2009)
7. N. Lourenço, N. Horta, GENOM-POF: multi-objective evolutionary synthesis of analog ics with corners validation, in *Proceedings of Genetic and Evolutionary Computation Conference (GECCO)*, (July 2012)

Chapter 7
Conclusions and Future Work

Abstract The proposed methodology for the automatic generation of analog integrated circuit (IC) layouts was proved by the implementation of a tool, LAYGEN II, which is able to generate robust layout solutions. This chapter presents the closing remarks, and the future directions for the continuous development of LAYGEN II.

Keywords Analog IC design · Automatic layout generation · Computer-aided-design · Electronic design automation

7.1 Conclusions

LAYGEN II is a combination of template-based and optimization-based approach, allowing the designer to provide layout guidelines that are used as a first cut solution allowing an intelligent pruning of the design space and, therefore, reducing the overall computational effort required by the evolutionary optimization kernel. Moreover, the use of a technology independent template, that creates an abstraction level between physical representation and designer's knowledge, introduce flexibility to the high level guidelines and eases the migration of designs to different IC technologies. LAYGEN II outstands from the remaining tools presented in the state-of-the-art on analog design automation of Chap. 2, by automatically generating flexible routing solutions, using only connectivity, that are validated in a commercial tool widely accepted in the industry, Calibre® [1] design rule check (DRC).

The tool potential has been proved for the two test cases presented. The hierarchical and modular nature of the developed approach allows the generation of large circuits layout by scaling the problem into different sub-templates, thus, dividing the problem into smaller ones. The introduction of an automatic generation during routing, independently from the floorplan attained, allows the designer to explore different layout topologies without the effort of defining new templates. The small computational times achieved for each automatic generation, reinforce the integration of the tool in the bottom-up physical synthesis path of an automatic analog design flow.

As this implementation stage, it is unlikely that automatically generated layouts would achieve better performance than handmade layouts. This limits the usage of the automatic tool in cells where the performance is extremely high, but for simpler cells or macro-cell place and route, this approach presents a highly effective design flow. The generated target layout must pass DRC and layout versus schematic (LVS) validations, and should not introduce extreme parasitic effects. While the parasitic are not handled at layout level, they are fed to an automatic circuit synthesizer that will re-size the circuit components to compensate layout parasitics, closing this way the traditional analog IC design flow.

Analog IC layout design automation is not a trivial matter and although LAYGEN II can automatically generate layouts, there is still much to evolve. This implementation focused on first settling an industrial-level layout synthesis process, and do not promptly take into account all precision parameters characteristics of the physical designs. Some limitations of the current implementation and relevant future improvements were identified. In the next section, these future enhancements are discussed and some concrete solutions presented.

7.2 Future Work

Starting with LAYGEN II's main purpose of creating an abstraction level between designer and technology details, the introduction of other automated abstraction levels above may increase the design automation level. The current implementation is gradually moving away from the template-based approach, but never tacking from the designer the ability to control the automatic generation.

The introduction of routing spaces constraints during placement reduce routing limitations that could arise from placement, however the generation flow placement-then-routing, still lacks the introduction of routing criteria during placement. Sometimes the minimum distances allowed by target technologies to place the devices may not be enough to router many wires. The use of a procedure which includes some measures to estimate the placement influence in routing may be considered in a future implementation.

As mentioned, a template developed for a current set of devices' sizes may not yield the best layout if those devices are changed. The currently NSGA-II algorithm [2] used opens a range of possible different implementations, which the most interesting is undoubtedly the extension of the placer module to allow topology exploration rather than strict template based generation. The designer guidelines should continue to be respected, such as symmetry, matching and proximity, but

7.2 Future Work

Fig. 7.1 Example of a pareto optimal front of placements

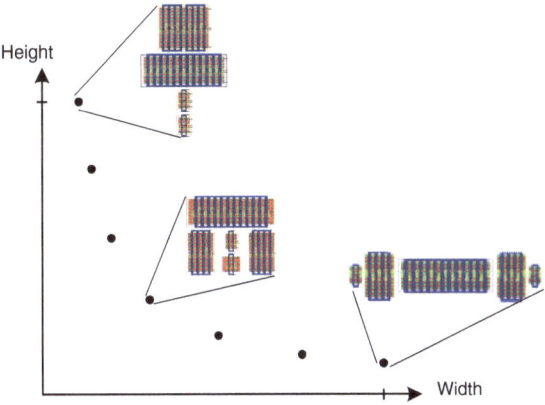

obtaining a Pareto optimal front (POF) of possible placements, instead of a single solution drawn directly from the topological relations between cells. An example of the possible POF generated is presented on Fig. 7.1. Since the routing connectivity and guidelines are provided independently from the obtained floorplan, the routing can be easily generated for each one of the solutions without requiring more effort by the designer.

The current B*-Tree layout representation although fits relatively well in the current template-based approach, for the topology exploration should be changed to a BSG or TSG-S representation. The possibility for the optimizer to know the relation between any pair of two cells is useful to set more compact floorplans, instead of the rectangular nature of the B*-Tree. It was demonstrated in Chap. 6 that the computational times of placer are not of concern.

Since it is desired to perform routing to each solution of the presented POF, an advanced version of the Router could identify patterns from the previous obtained solutions and use them as the starting point for the following evolutionary algorithms. This could mean a relative improvement on the computation time, since only the optimization phases of the first floorplans routed would take the total execution time, but the following routing tasks would converge faster having a previous optimized solution as the starting point. Theoretically, this would improve significantly the routing quality of the solutions achieved. Since every solution of the POF is composed of the same devices, the topological relation patterns are common between the floorplan solutions.

The parametric module generator of technology design kits should also be extended to interdigitized and common centroid cells. The processing of matched cells is already implemented, only the specific parametric cells are missing. In placement processing, the convex hull approach used for guard ring processing may be adapted for creating unique wells that contain different transistors. At this moment, if the respective wells from transistors are packed together the minimum distances processing do not depart them, but the lack of a unique well for those transistors that do not join was noticed.

For router, the lack of 45° wires was identified and should be implemented in future versions of LAYGEN II. The actual wire structure and connectivity supports 45° segments, as genetic operators can be easily adapted to chromosomes containing these wires. However, the main obstacle is the internal evaluation procedure. The actual implementation of the internal evaluator uses the bounding boxes of shapes contained in the layout to verify the intersections and distances, and these bounding boxes are only rectangles. The introduction of non 90° segments in the layout force different geometrical considerations, and computing distances or intersections must be converted to a point-to-point approach, instead of the single coordinates approach.

The endnotes of future developments shall be the improvements required to LAYGEN II to be compliant with deep nanometer technologies. The inclusions of transistor dummies in the current modules and to consider more specific ERC, e.g., well proximity effect or time-dependent dielectric breakdown, are just some examples of the challenges that have to be considered as it is being dealt with smaller design processes. Although being applicable to any design process, they become more relevant for the process variability of the sub-100 nm ICs era. While still under development, LAYGEN II has proven capable to assist the designer to obtain a robust first cut design and so is intended to maintain competitive for the deep nanometer processes.

The use of optimization techniques for analog ICs layout generation has been demonstrated, however, the potential for further developments is still large, some of these were identified implementations for the recent future. The LAYGEN II project is not closed, far from it, this work served to validate the concept and it provides support for further developments. New people will bring new insights in the analog IC layout generation problem, and it is hoped that LAYGEN II will converge to an application suitable for industrial uses, it is being developed to be embedded in an automatic analog IC design flow following the indications of the recent published results, suggesting that soon a viable unified sizing/layout solutions shall arise and settle in the electronic design automation market.

References

1. Mentor Graphics, http://www.mentor.com
2. K. Deb, A. Pratap, S. Agarwal, T. Meyarivan, A fast and elitist multiobjective genetic algorithm: NSGA-II. IEEE Trans. Evol. Comput. **6**(2), 182–197 (2002)